Audel™

Carpenters and Builders Millwork, Power Tools, and Painting

All New 7th Edition

Audel™

Carpenters and Builders Millwork, Power Tools, and Painting

All New 7th Edition

Mark Richard Miller
Rex Miller

Wiley Publishing, Inc.

Vice President and Executive Group Publisher: Richard Swadley
Vice President and Executive Publisher: Robert Ipsen
Vice President and Publisher: Joseph B. Wikert
Executive Editorial Director: Mary Bednarek
Executive Editor: Carol Long
Editorial Manager: Kathryn A. Malm
Senior Production Editor: Geraldine Fahey
Development Editor: Kevin Shafer
Production Editor: Pamela Hanley
Text Design & Composition: TechBooks

Published simultaneously in Canada

For general information on our other products and services please contact our Customer Care Department within the United States at (800) 762-2974, outside the United States at (317) 572-3993 or fax (317) 572-4002.

Library of Congress Cataloging-in-Publication Data:

ISBN: 0-764-57114-1

Printed in the United States of America

10 9 8 7 6 5 4 3 2 1

Contents

Acknowledgments

No book can be written without the aid of many people. It takes a great number of individuals to put together the information available about any particular technical field into a book. Many firms have contributed information, illustrations, and analyses of the book.

The authors would like to thank every person involved for his or her contributions. Following are some of the firms that supplied technical information and illustrations:

American Plywood Association

Belsaw

Benjamin Moore

Certain-Teed

Congoleum

Dow

Georgia-Pacific Corp.

Kitchen & Bath Source Book

Mineral Wool Manufacturing Association

Olean Tile

Owens-Corning

Rockwell

Sears, Roebuck, and Co.

Shopsmith

Stanley Tools

United Gilsonite

About the Authors

Rex Miller was a Professor of Industrial Technology at The State University of New York, College at Buffalo for more than 35 years. He has taught on the technical school, high school, and college level for more than 40 years. He is the author or coauthor of more than 100 textbooks ranging from electronics through carpentry and sheet metal work. He has contributed more than 50 magazine articles over the years to technical publications. He is also the author of seven Civil War regimental histories.

Mark Richard Miller finished his BS degree in New York and moved on to Ball State University, where he obtained the Masters and went to work in San Antonio. He taught in high school and went to graduate school in College Station, Texas, finishing the doctorate. He took a position at Texas A&M University in Kingsville, Texas, where he now teaches in the Industrial Technology Department as a Professor and Department Chairman. He has coauthored 11 books and contributed many articles to technical magazines. His hobbies include refinishing a 1970 Plymouth Super Bird and a 1971 Road-Runner.

Preface

The *Audel Carpenters and Builders Millwork, Power Tools, and Painting: All New Seventh Edition* is the fourth of four volumes that cover the fundamental tools, methods, and materials used in carpentry, woodworking, and cabinetmaking.

This volume is written for anyone who wants (or needs) to understand how to use power tools, do millwork, or paint a building (or its interior). The problems encountered here can make or break a house or any project. Problems encountered by the carpenter, woodworker, cabinetmaker or do-it-yourselfer often need attention by someone familiar with the requirements of a job done properly. Whether remodeling an existing home or building a new one, the rewards of doing a good job are great.

This book has been prepared for use as a practical guide in the selection, maintenance, installation, operation, and repair of wooden structures. Carpenters and woodworkers (as well as cabinetmakers and new homeowners) should find this book, with its clear descriptions, illustrations, and simplified explanations, a ready source of information for the many problems that they might encounter while building, maintaining, or repairing houses and or furniture. Both professionals and do-it-yourselfers who want to gain knowledge of woodworking and house building will benefit from the theoretical and practical coverage of this book.

This is the fourth of a series of four books in the Carpenters and Builders Library designed to provide you with a solid reference set of materials that can be useful both at home and in the field. Other books in the series include the following:

- *Audel Carpenters and Builders Tools, Steel Square, and Joinery: All New Seventh Edition*
- *Audel Carpenters and Builders Math, Plans, and Specifications: All New Seventh Edition*
- *Audel Carpenters and Builders Millwork, Layout, Foundation, and Framing: All New Seventh Edition*

No book can be completed without the aid of many people. The Acknowledgments mention some of those who contributed to making this the most current in design and technology available to the carpenter. We trust you will enjoy using the book as much as we did writing it.

Mark R. Miller
Rex Miller

Chapter 1

Stairs

Anyone who has tried to build stairs has found it to be an art in itself. This chapter is not intended to discourage the builder, but rather to impress upon the builder the fact that unless he first masters the principle of stair layout, he will have many difficulties in the construction. Although stair building is a branch of millwork, the carpenter should know the principles of simple stair layout and construction because of often being called upon to construct porch steps, basement and attic stairs, and other stairs. To follow the instructions intelligently, the carpenter should be familiar with the terms and names of parts used in stair building.

Stair Construction

Stairways should be designed, arranged, and installed to afford safety, adequate headroom, and space for the passage of furniture. In general, there are two types of stairs in a house: those serving as principal stairs and those used as service stairs. The principal stairs are designed to provide ease and comfort and are often made a feature of design. The service stairs leading to the basement or attic are usually somewhat steeper and constructed of less-expensive materials.

Stairs may be built in place, or they may be built as units in the shop and set in place. Both methods have their advantages and disadvantages, and custom varies with locality. Stairways may have a straight continuous run, with or without an intermediate platform, or they may consist of two or more runs at angles to each other. In the best and safest practice, a platform is introduced at the angles, and the radiating risers are called winders. Winder stairways are hazardous.

Ratio of Riser to Tread

There is a definite relationship between the height of a riser and the width of a tread, and all stairs should be laid out to conform to the well-established rules governing this relationship. If the combination of run and rise is too great, the steps are tiring, placing a strain on the leg muscles and on the heart. If the steps are too short, the foot may kick the leg riser at each step, and an attempt to shorten the stride may be tiring. Experience has proved that a riser 7 to $7^{1}/_{2}$ inches high, with appropriate tread, combines both comfort and safety, and these limits therefore determine the standard height of risers commonly used for principal stairs. Service stairs may be narrower and steeper than the principal stairs, and are often unduly so, but it is a good idea not to exceed 8 inches for the risers.

As the height of the riser is increased, the width of the tread must be decreased for comfortable results. A very good ratio is provided by either of the following rules, which are exclusive of the nosing:

- Tread plus twice the riser equals 25.
- Tread multiplied by the riser equals 75.

A riser of $7^1/_2$ inches therefore, require a tread of 10 inches, and a riser of $6^1/_2$ inches would require a tread 12 inches wide. Treads are rarely made less than 9 inches or more than 12 inches wide. The treads of main stairs should be made of prefinished hardwood. Following are some riser and tread rules to keep in mind:

- The actual riser and tread dimensions for a set of stairs are determined by dividing the total rise (or floor-to-floor height) by the desired riser height. The result is rounded off to arrive at a whole number of risers. The total rise is then redivided by this whole number to arrive at the actual riser height.
- This riser height must be checked against the maximum riser height allowed by the building code. If necessary, the number risers can be increased by one and the actual riser height recalculated.
- Once the actual riser height is fixed, the tread run can be determined by using the riser-to-tread proportioning formula.
- Since in any flight of stairs there is always one less tread than the number of risers, the total number of treads and the total run can be easily determined (Table 1-1).

Design of Stairs

The location and the width of a stairway (together with the platforms) having been determined, the next step is to fix the height of the riser and width of the tread. After a suitable height of riser is chosen, the exact distance between the finish floors of the two stories under consideration is divided by the riser height. If the answer is an even number, the number of risers is thereby determined. It very often happens that the result is uneven, in which case the story height is divided by the whole number next above or below the quotient. The result of this division gives the height of the riser. The tread is then proportioned by one of the rules for ratio of riser to tread.

Table 1-1 Measurement of Riser and Tread

Riser Inches (mm)	*Tread Inches (mm)*
5 (125)	15 (380)
$5^1/_4$ (135)	$14^1/_2$ (370)
$5^1/_2$ (140)	14 (355)
$5^3/_4$ (145)	$13^1/_2$ (340)
6 (150)	13 (330)
$6^1/_4$ (160)	$12^1/_2$ (320)
$6^1/_2$ (165)	12 (305)
$6^3/_4$ (170)	$11^1/_2$ (290)
7 (180)	11 (280)
$7^1/_4$ (185)*	$10^1/_2$ (265)*
$7^1/_2$ (190)*	10 (255)*
$7^3/_4$ (195)*	$9^1/_2$ (240)*
8 (205)*	9 (230)*

*These riser and tread dimensions are permitted only for private stairways serving an occupancy of less than 10 and stairways leading to an unoccupied roof.

For example, assume that the total height from one floor to the top of the next floor is 9 feet 6 inches, or 114 inches, and that the riser is to be approximately $7^1/_2$ inches. The 114 inches would be divided by $7^1/_2$ inches, which would give $15^1/_5$ risers. However, the number of risers must be an equal or whole number. Since the nearest whole number is 15, it may be assumed that there are to be 15 risers, in which case 114 divided by 15 equals 7.6 inches. That is approximately $7^9/_{16}$ inches for the height of each riser. To determine the width of the tread, multiply the height of the riser by 2 (that is, $2 \times 7^9/_{16} = 15^1/_8$), and deduct from 25 (that is, $25 - 15^1/_8 = 9^7/_8$ inches).

The headroom is the vertical distance from the top of the tread to the underside of the flight or ceiling above (Figure 1-1). Although it varies with the steepness of the stairs, the minimum allowed would be 6 feet 8 inches.

Framing of Stairwell

When large openings are made in the floor (such as for a stairwell), one or more joists must be cut. The location in the floor has a direct bearing on the method of framing the joists.

The principles explained in Chapter 7 of the third book in this series, *Audel Carpenters and Builders Layout, Foundation, and*

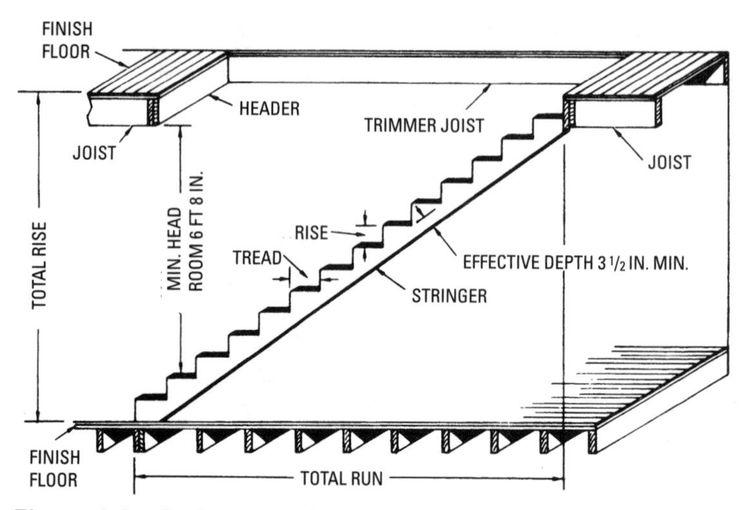

Figure 1-1 Stairway design.

Framing (Hoboken, N.J.: Wiley Publishing, 2005) may be referred to in considering the framing around openings in floors for stairways. (See the Preface for more information on the series.) The framing members around these openings are generally of the same depth as the joists (Figure 1-2).

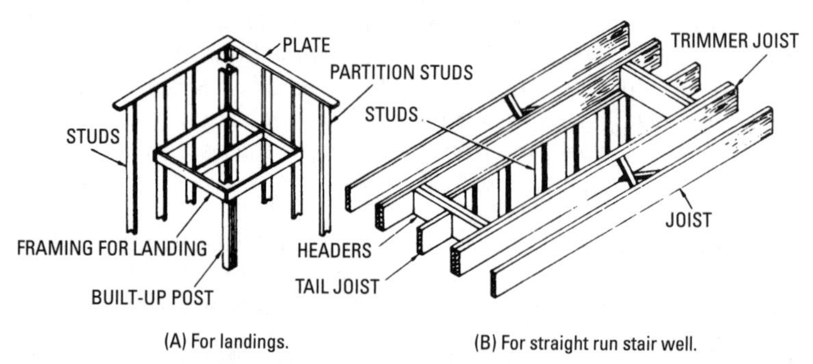

Figure 1-2 Framing of stairways.

The headers are the short beams at right angles to the regular joists at the end of the floor opening. They are doubled. They support the ends of the joists that have been cut off. Trimmer joists are at the sides of the floor opening and run parallel to the regular joists. They are doubled. And they are used to support the ends of the headers. Tail joists are joists that run from the headers to the bearing position.

Stringers or Carriages

The treads and risers are supported upon stringers, or carriages that are solidly fixed in place and are level and true on the framework of the building. The stringers may be cut or ploughed to fit the outline of the tread and risers. The third stringer should be installed in the middle of the stairs when the treads are less than $1^1/_8$ inches thick and the stairs are more than 2 feet 2 inches wide. In some cases, rough stringers are used during the construction period. These have rough treads nailed across the stringers for the convenience of workers until the wall finish is applied. There are several forms of stringers classed according to the method of attaching the risers and treads. These different types are cleated, cut, built-up, and rabbeted.

When the wall finish is complete, the finish stairs are erected or built in place. This work is generally done by a stair builder (who is a specialist), but the carpenter also does the job. The wall stringer may be ploughed out, or rabbeted (Figure 1-3), to the exact profile

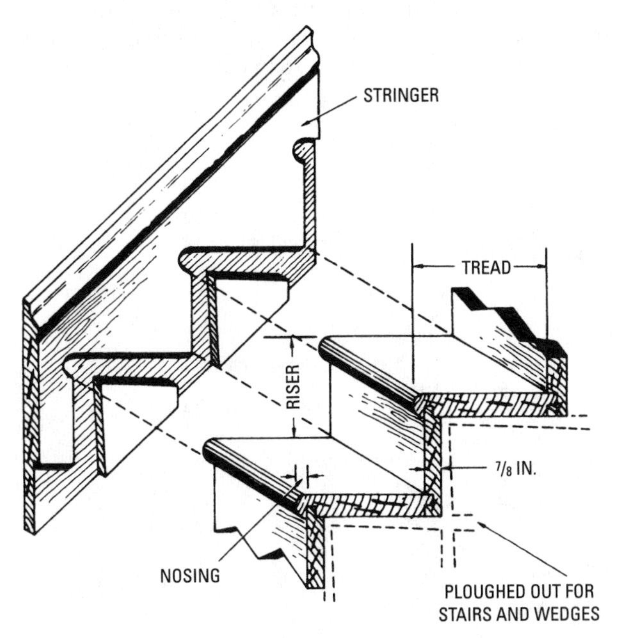

Figure 1-3 The housing in the stringer board for the tread and riser.

of the tread, riser, and nosing, with sufficient space at the back to take the wedges. The top of the riser is tongued into the front of the tread and into the bottom of the next riser. The wall stringer is spiked to the inside of the wall. The treads and risers are fitted together and forced into the wall stringer nosing, where they are set tight by driving and gluing the wood wedges behind them. The wall stringer shows above the profiles of the tread and riser as a finish against the wall, and is often made continuous with the baseboard of the upper and lower landing. If the outside stringer is an open stringer, it is cut out to fit the risers and treads and nailed against the outside carriage. The edges of the riser are mitered with the corresponding edges of the stringer, and the nosing of the tread is returned upon its outside edge along the face of the stringer. Another method would be to butt the stringer to the riser and cover the joint with an inexpensive stair bracket.

Figure 1-4 shows a finish stringer nailed in position on the wall and the rough carriage nailed in place against the stringer. If there are walls on both sides of the staircase, the other stringer and carriage would be located in the same way. The risers are nailed to the riser cuts of the carriage on each side and butt against each side of the stringer. The treads are nailed to the tread cuts of the carriage and butt against the stringer. This is the least expensive of the types described and perhaps the best construction to use when the treads and risers are to be nailed to the carriages.

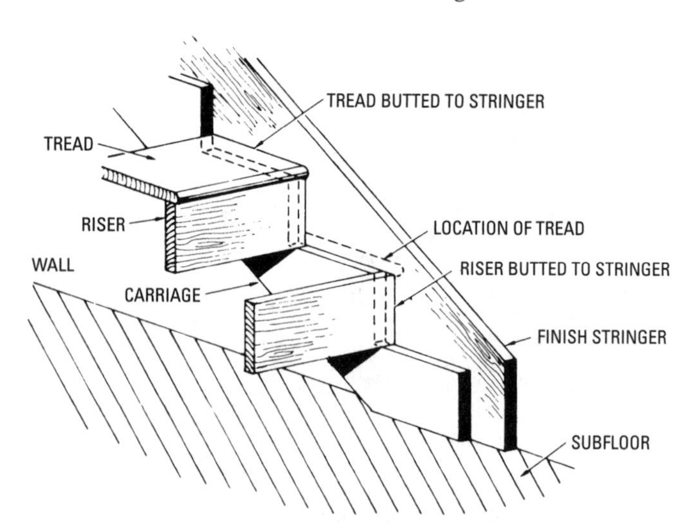

Figure 1-4 Finished wall stringer and carriage.

Figure 1-5A shows another method of fitting the treads and risers to the wall stringers. The stringers are laid out with the same rise and run as the stair carriages, but they are cut out in reverse. The risers are butted and nailed to the riser cuts of the wall stringers, and the assembled stringers and risers are laid over the carriage. Sometimes the treads are allowed to run underneath the tread cut of the stringer. This makes it necessary to notch the tread at the nosing to fit around the stringer (Figure 1-5B).

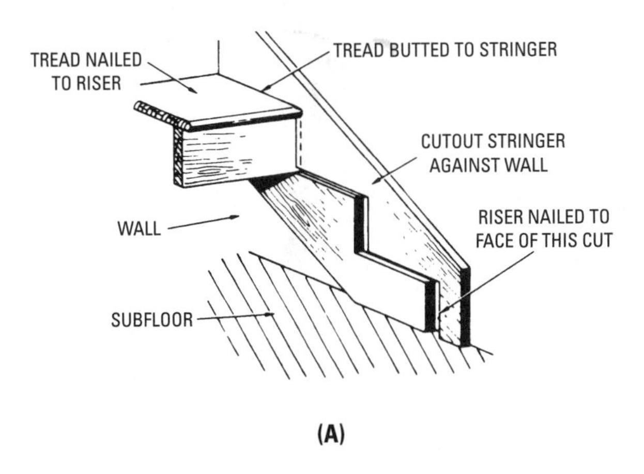

(A)

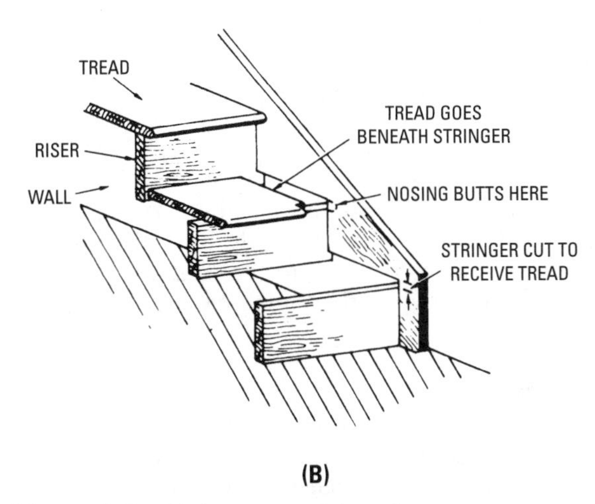

(B)

Figure 1-5 Stringers and treads.

Another form of stringer is the cut-and-mitered type. This is a form of open stringer. The ends of the risers are mitered against the vertical portion of the stringer. This construction is used when the outside stringer is to be finished and must blend with the rest of the casing or skirting board (Figure 1-6). A molding is installed on the edge of the tread and carried around to the side, making an overlap (Figure 1-7).

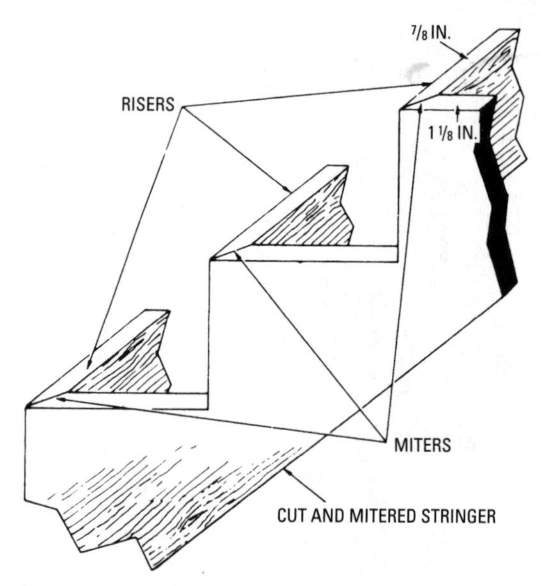

Figure 1-6 Cut-and-mitered stringer.

Basement Stairs

Basement stairs may be built either with or without riser boards. Cutout stringers are probably the most widely used support for the treads, but the tread may be fastened to the stringers by cleats (Figure 1-8). Figure 1-9 shows two methods of terminating basement stairs at the floor line.

Newels and Handrails

All stairways should have a handrail from floor to floor. For closed stairways, the rail is attached to the wall with suitable metal brackets. The rails should be set 2 feet 8 inches above the tread at the riser line. Handrails and balusters are used for open stairs and for

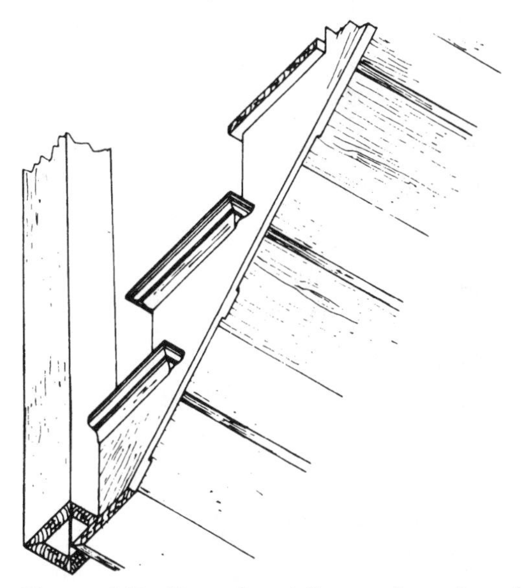

Figure 1-7 Use of molding at the edge of the tread.

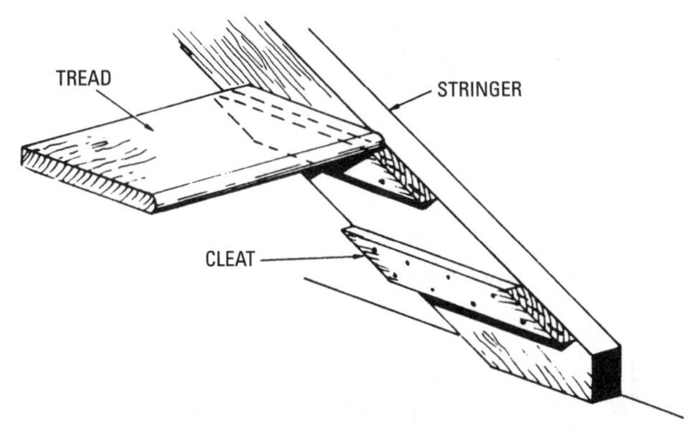

Figure 1-8 Cleat stringer.

open spaces around stairs. The handrail ends against the newel post (Figure 1-10).

Stairs should be laid out so that stock parts may be used for newels, rails, balusters, goosenecks, and turnouts. These parts are a matter of design and appearance, so they may be very plain or elaborate, but they should be in keeping with the style of the house. The balusters are doweled or dovetailed into the treads and, in some

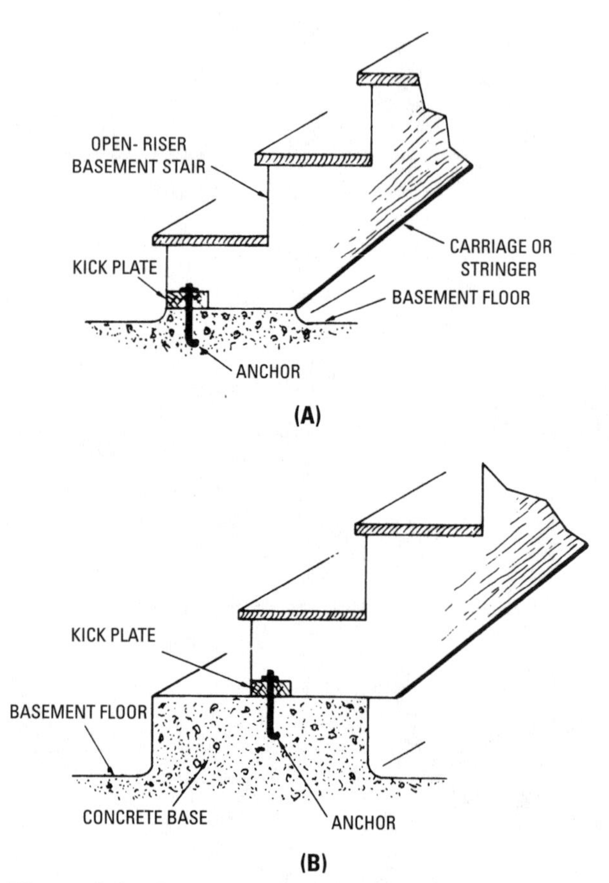

Figure 1-9 Basement stair at floor level.

cases, are covered by a return nosing. Newel posts should be firmly anchored, and where half-newels are attached to a wall, blocking should be provided at the time the wall is framed.

Disappearing Stairs

Where attics are used primarily for storage, and where space for a fixed stairway is not available, hinged or disappearing stairs are often used. Such stairways may be purchased ready to install. They operate through an opening in the ceiling of a hall and swing up into the attic space, out of the way, when not in use. Where such stairs

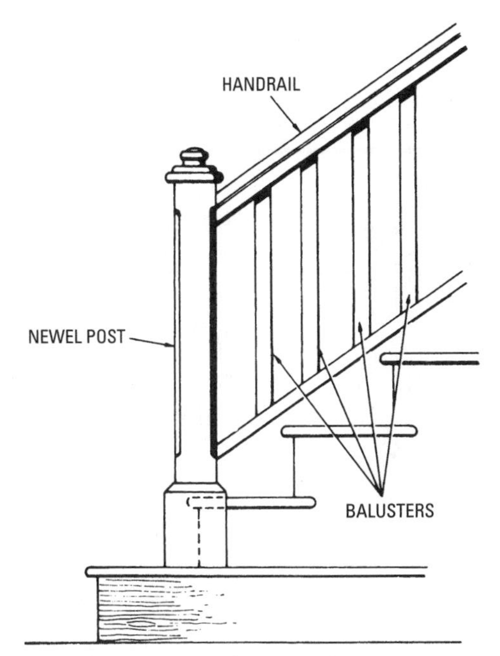

Figure 1-10 Illustrating the newel post, balusters, and handrail.

are to be provided, the attic floor should be designed for regular floor loading.

Exterior Stairs

Proportioning of risers and treads in laying out porch steps or approaches to terraces should be as carefully considered as the design of interior stairways. Similar riser-to-tread ratios can be used, however. The riser used in principal exterior steps should be between 6 and 7 inches. The need for a good support or foundation for outside steps is often overlooked. Where wood steps are used, the bottom step should be set in concrete. Where the steps are located over backfill or disturbed ground, the foundation should be carried down to undisturbed ground. Figure 1-11 shows the foundation and details of step treads, handrail, and stringer, and the method of installing them. This type of step is most common in outside porch steps. The material generally used for this type of stair construction is pressure-treated 2 × 4 boards and 2 × 6 boards.

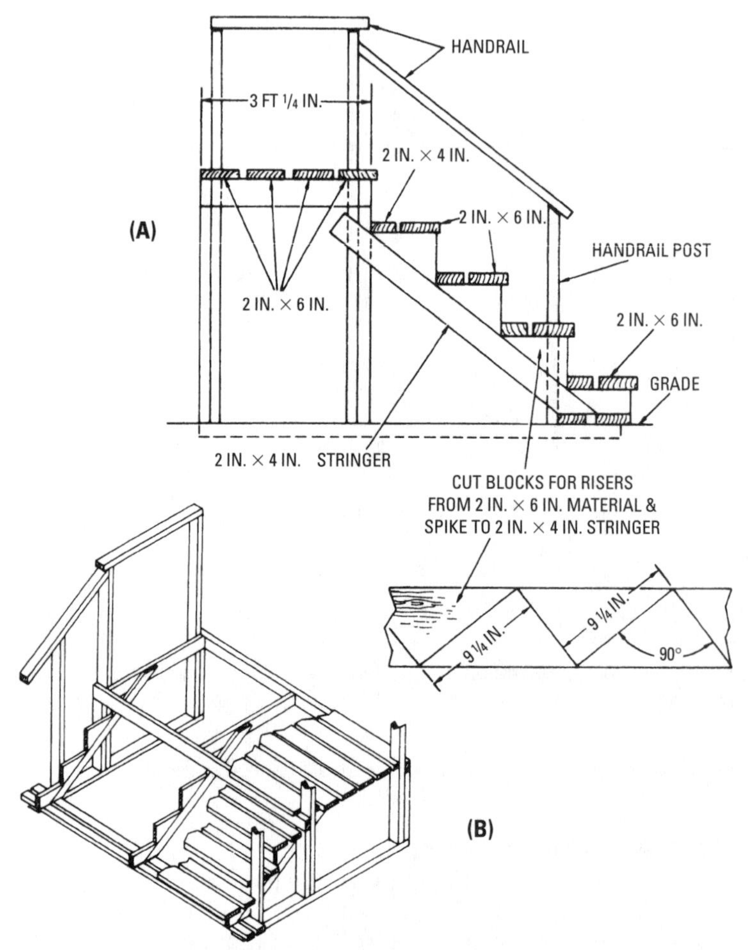

Figure 1-11 Outside step construction.

Glossary of Stair Terms

The terms generally used in stair design may be defined as follows:

- Balusters—The vertical members supporting the handrail on open stairs (Figure 1-12).
- Carriage—The rough timber supporting the treads and risers of wood stairs is sometimes referred to as the string or stringer (Figure 1-13).

Figure 1-12 Vertical balusters support the handrail.

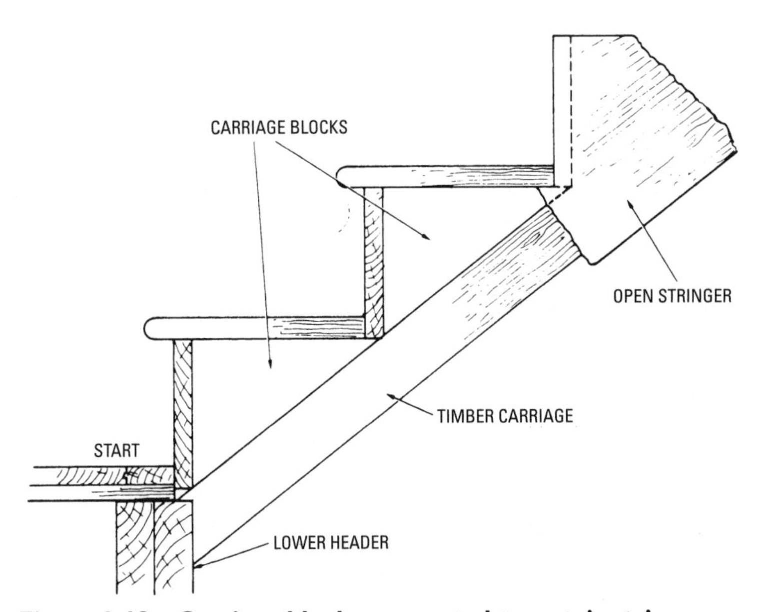

Figure 1-13 Carriage blocks connected to a stair stringer.

- Circular stairs—A staircase with steps planned in a circle, all the steps being winders (Figure 1-14).
- Fillet—A band nailed to the face of a front string below the curve and extending the width of a tread.
- Flight of stairs—The series of steps leading from one landing to another.

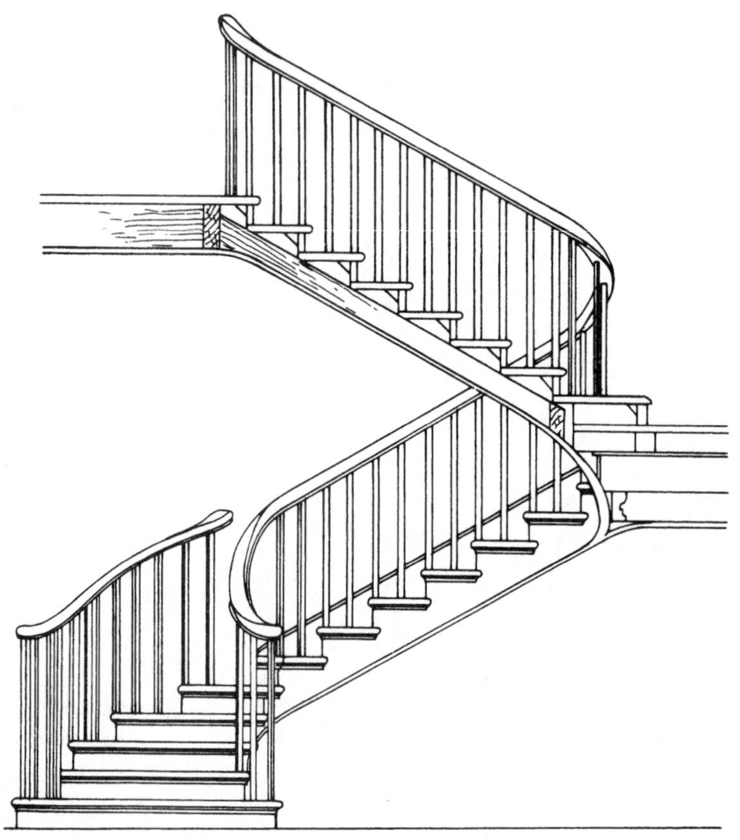

Figure 1-14 A staircase with winding treads.

- Flyers—Steps in a flight of stairs parallel to each other.
- Front string or stringer—The stringer on that side of the stairs over which the handrail is placed.
- Half-space—The interval between two flights of steps in a staircase.
- Handrail—The top finishing piece on the railing intended to be grasped by the hand in ascending and descending. For closed stairs where there is no railing, the handrail is attached to the wall with brackets (Figure 1-15).
- Housing—The notches in the stringboard of a stair for the reception of steps.

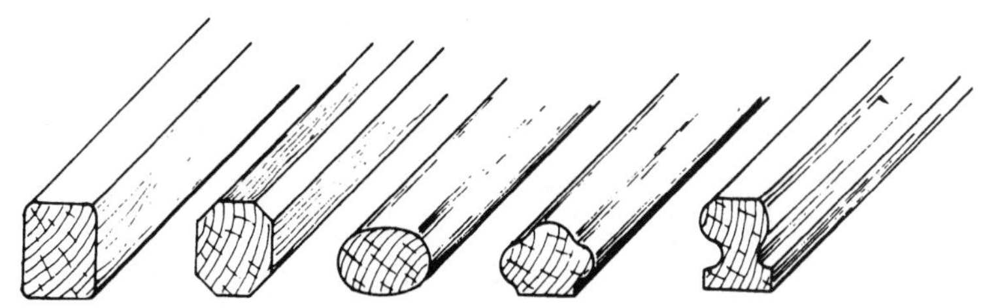

Figure 1-15 Various profiles of handrails.

- Landing—The floor at the top or bottom of each story where the flight ends or begins.
- Newel—The main post of the railing at the start of the stairs and the stiffening posts at the angles and platform.
- Nosing—The projection of tread beyond the face of the riser (Figure 1-16).
- Rise—The vertical distance between the treads or for the entire stairs.
- Riser—The board forming the vertical portion of the front of the step (Figure 1-17).
- Run—The total length of stairs including the platform.

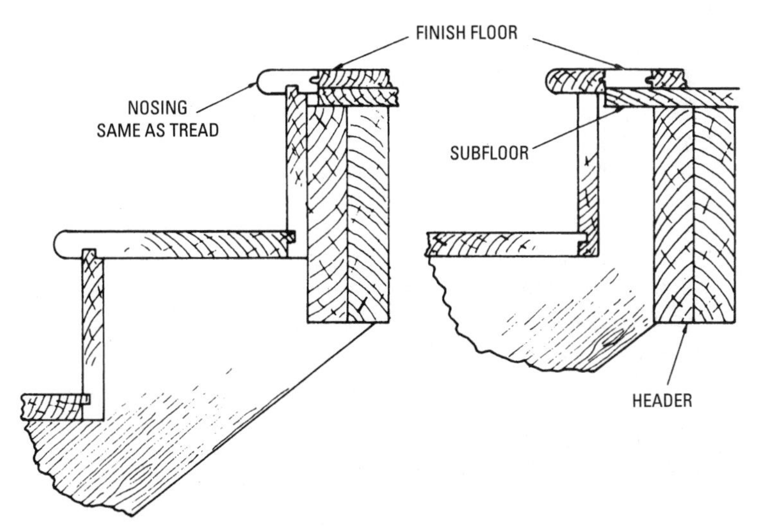

Figure 1-16 The nosing installed to the tread.

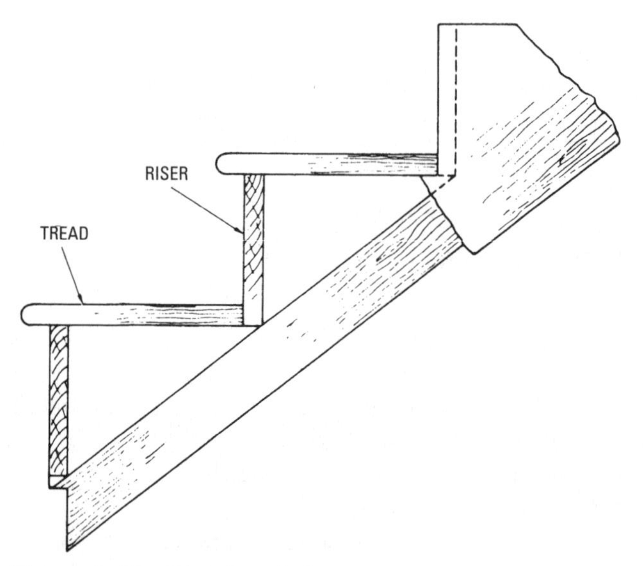

Figure 1-17 Tread and riser.

- Staircase—The whole set of stairs with the side members supporting the steps.
- Stairs—The steps used to ascend and descend from one story to another.
- Straight flight of stairs—Flight of stairs in which the steps are parallel and at right angles to the strings.
- String or stringer—One of the inclined sides of a stair that supports the tread and riser. Also, a similar member (whether a support or not) such as finish stock placed exterior to the carriage on open stairs, and next to the walls on closed stairs, to give finish to the staircase. Open stringers, both rough and finish stock, are cut to follow the lines of the treads and risers. Closed stringers have parallel sides, with the risers and treads being housed into them (Figure 1-18).
- Tread—The horizontal face of a step, as shown in Figure 1-17.
- Winders—The radiating or wedge-shaped treads at the turn of a stairway.

Building a Curved Stairway

There are a number of unique problems associated with the building of a curved stairway. One of the first thoughts is, "How do I get a handrail to fit the curve of this stairway?"

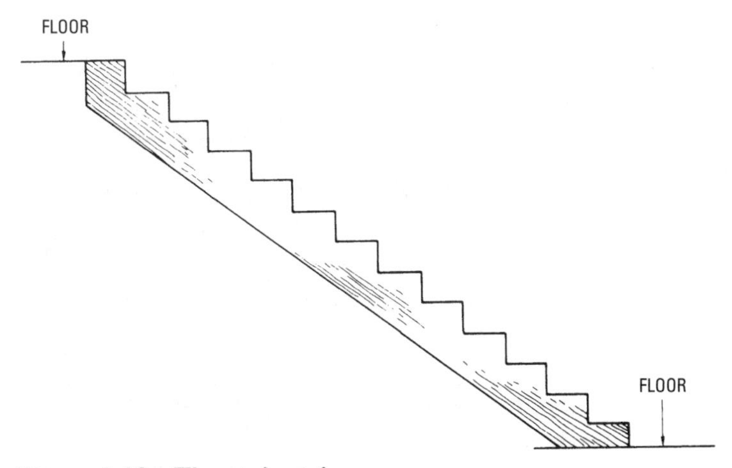

Figure 1-18 The stair stringer.

In the next series of pictures, you will see how this is done by a very patient carpenter who specializes in this type of problem. This house is built on a slab and there is a second story with an open family room on the first floor leading to a game room on the second. The circular stairway improves the first impression

Figure 1-19 Most of the house has been framed and the stairs are in place.

Figure 1-20 Note the curved portion of the stair case attached to the floor.

Figure 1-21 Close up view of the prefabricated last step coming down.

Figure 1-22 Stairway enclosed with drywall and step at bottom being installed.

Figure 1-23 First strips being held by clamps from the woodshop.

Figure 1-24 Clamps starting the process upstairs.

of the foyer and the house as a whole and makes it easy for a builder to sell this type of house once completed and used as a model.

The first step is a prefabricated piece of oak designed for hard use as a landing for all those coming down the stairs. The other steps will be carpeted so no special attention will have to be paid to the treads and risers since they will be covered. Figure 1-19 shows the framing of the area where the stairway will be located. Figure 1-20 shows how the curve has been established on the floor and anchored into the concrete. Figure 1-21 is a closer look at the bottom step before the tread is applied or glued and screwed in place. Note the two bottles of glue that the carpenter will need to bend the handrails to fit the stairway. Figure 1-22 shows the curvature of the staircase.

The first step in bending the handrail begins upstairs where a couple of the strips of oak that are thin enough to bend easily are glued together and clamped to conform to the curve of the stairway (Figure 1-23). Later, more strips will be added, and each time the added strip will be clamped until the glue dries. You can see from this procedure that it takes some time for the job to be done properly.

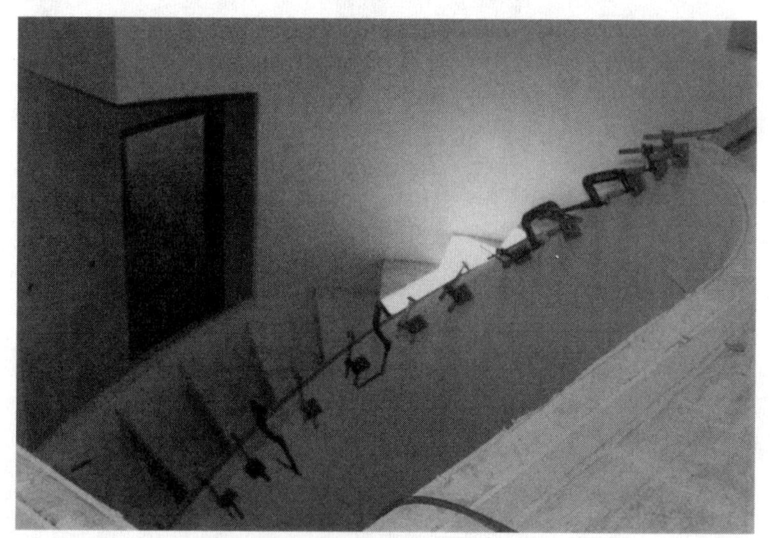

Figure 1-25 Clamps on the longest run of the gluing process.

Figure 1-24 and Figure 1-25 show clamps and oak strips being bent to conform to the stairway. Figure 1-26 shows how two examples of the wood strips and their molds are assembled to make the handrail when the glue and the process have been completed. The outside pieces of molding strip prevent damage to the oak strips used for the handrail.

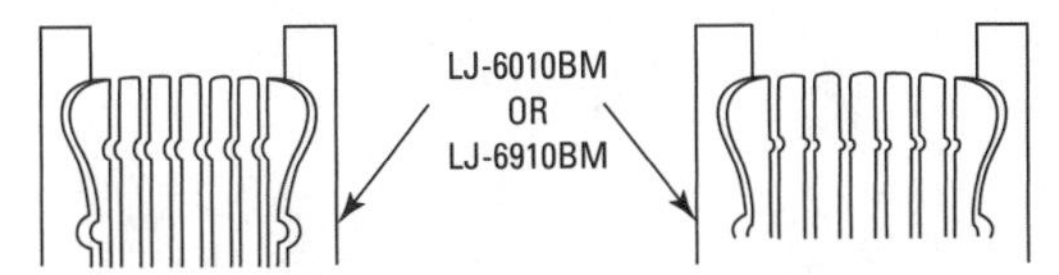

Figure 1-26 Molds that hold the strips of oak being glued.

Figure 1-27 and Figure 1-28 show more of the bending process of making a handrail fit its stairway. Note in Figure 1-29 the handrails have been installed and balusters have begun to take their place in the railing. Figure 1-30, Figure 1-31, and Figure 1-32 show the posts being installed before the hand railing can be attached. Figure 1-33 shows the upstairs balusters installed. Figure 1-34 shows the staircase being utilized as part of the furniture.

Figure 1-27 Clamps on the other and lower side of the staircase.

Figure 1-28 Clamps holding the glued strips so they form the same curvature as the staircase.

Figure 1-29 Handrails installed and installation of balusters progressing.

Figure 1-30 Starting the installation of posts.

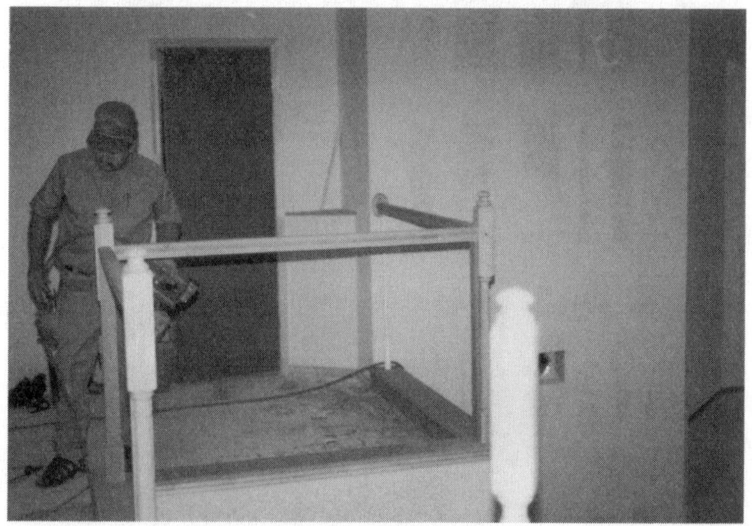

Figure 1-31 First installation of the handrail.

Figure 1-32 Posts being installed.

Figure 1-33 Handrail and balusters in place.

Figure 1-34 Finished staircase.

Summary

Stairways should always be designed, arranged, and installed for safety, adequate headroom, and space for passage of furniture. Stairs may be built in place, or they may be built as a complete unit in the shop and set in place.

There is a definite relationship between the height of a riser and the width of a tread. If the steps are too short, the foot may hit the leg riser at each step and an attempt to shorten the stride may be tiring. As the height of the riser is increased, the width of the tread must be decreased for comfortable use.

When openings are made in a floor (such as for a stairwell), headers and trimmer joists must be used to strengthen the floor around the opening. Stringers (which are the supports for the tread and riser) are installed between floor levels. There are several forms of stringers classed according to the method of attaching the risers and treads. The different types are cleated, cut, built-up, and rabbeted.

Review Questions

1. What are stringers, risers, and treads?
2. Name the four types of stringers.
3. When is a center or third stringer used?
4. How is the rise figured when designing a stairway?
5. What is the carriage of a staircase?

Chapter 2

Flooring

After the foundation, sills, and floor joists have been constructed, the subfloor is laid on the joists. The floor joist forms a framework for the subfloor. This floor is called the *rough floor* (or *subfloor*), and may be viewed as a large platform covering the entire width and length of the building. Two layers or coverings of flooring material (subflooring and finished flooring) are placed on the joists. You may use 1 × 4 boards, 1 × 6 tongue-and-groove sheathing, or $^5/_8$-inch plywood for the subfloor.

Floor Coverings

There is a wide variety of finish flooring available, each having properties suited to a particular usage. Of these properties, durability and ease of cleaning are essential. Specific service requirements may call for special properties, such as

- Resistance to hard wear in storehouses and on loading platforms
- Comfort to users in offices and shops
- Attractive appearance, which is always desirable in residences

Both hardwoods and softwoods are available as strip flooring in a variety of widths and thicknesses, as well as random-width planks, parquetry, and block flooring.

Wood Strip Flooring

Softwoods most commonly used for flooring are Southern yellow pine, Douglas fir, redwood, Western larch, and Western hemlock. It is customary to divide the softwoods into two classes:

- Vertical or edge-grain
- Flat grain

Each class is separated into select and common grades. The select grades designated as "B and better" grades (and sometimes the "C" grade) are used when the purpose is to stain, varnish, or wax the floor. The "C" grade is better suited for floors to be stained dark or painted, and lower grades are for rough usage when covered with carpeting. Softwood flooring is manufactured in several widths. In some regions, the $2^1/_2$-inch width is preferred, while in others the $3^1/_2$-inch width is more popular. Softwood flooring has tongue-and-groove edges and may be hollow-backed or grooved.

Vertical-grain flooring stands up better than flat-grain flooring under hard usage.

Hardwoods most commonly used for flooring are

- Red oak
- White oak
- Hard maple
- Beech
- Birch

Maple, beech, and birch come in several grades (such as *first, second*, and *third*). Other hardwoods that are manufactured into flooring, although not commonly used, are walnut, cherry, ash, hickory, pecan, sweet gum, and sycamore.

Hardwood flooring is manufactured in a variety of widths and thicknesses, some of which are referred to as *standard patterns*, others as *special patterns*. The widely used standard patterns consist of relatively narrow strips laid lengthwise in a room (Figure 2-1). The most widely used standard pattern is $^{25}/_{32}$-inch thick and has a face width of $2^1/_4$ inches. One edge has a tongue and the other end has a groove, and the ends are similarly matched. The strips are random lengths, varying from 1 to 16 feet in length. The number of short pieces will depend on the grade used. Similar patterns of flooring are available in thicknesses of $^{15}/_{32}$ and $^{11}/_{32}$ inch, with a face width of $1^1/_2$ inch, and with square edges and a flat back.

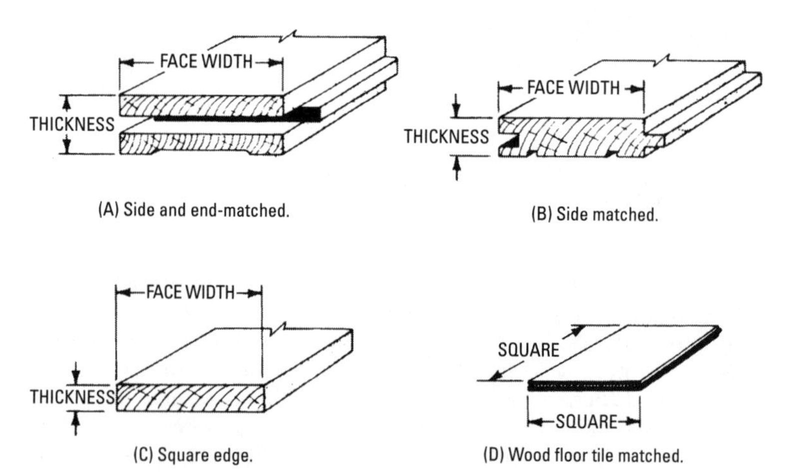

Figure 2-1 Types of finished hardwood flooring.

Flooring is generally hollow-backed. The top face is slightly wider than the bottom, so that the strips are driven tight together at the topside, but the bottom edges are slightly open. The tongue should fit snugly in the groove to eliminate squeaks in the floor.

Another pattern of flooring used to a limited degree is $^{3}/_{8}$-inch thick with a face width of $1^{1}/_{2}$ and 2 inches, with square edges and a flat back. Figure 2-1D shows a type of wood floor tile.

Installation of Wood Strip Flooring

Flooring should be laid after the wall and ceiling are completed, after windows and exterior doors are in place, and after most of the interior trim is installed. The subfloor should be clean and level, and should be covered with a 15-pound asphalt building paper (Figure 2-2). This building paper will stop a certain amount of dust and will somewhat deaden the sound. Where a crawl space is used, it will increase the warmth of the floor by preventing air infiltration. The location of the joists should be chalk-lined on the paper as a guide for nailing.

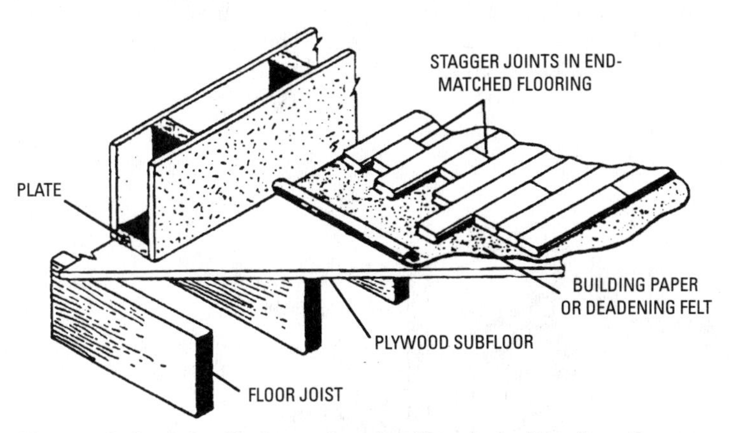

Figure 2-2 Installation of strip flooring showing the use of deadening felt or heavy building paper.

Strip flooring should be laid crosswise to the floor joists, and it looks best when the floor is laid lengthwise in a rectangular room. Since joists generally span the short way in a living room, that room establishes the direction for flooring in other rooms. Flooring should be delivered only during dry weather and should be stored in the warmest and driest place available in the house. The recommended average moisture content for flooring at the time of installation should be between 6 and 10 percent. Moisture absorbed after the material is delivered to the house site is one of the most common causes of open joints between floor strips. It will show up several months after the floor has been laid.

Floor squeaks are caused by the movement of one board against another. Such movement may occur because the floor joists are too light and are not held down tightly, the tongue fits too loosely in the grooves, or because of poor nailing. Adequate nailing is one of the most important means of minimizing squeaks. When it is possible to nail the finish floor through the subfloor into the joist, a much better job is obtained than if the finish floor is nailed only to the subfloor.

Various types of nails are used in nailing various thicknesses of flooring. For $^{25}/_{32}$-inch flooring, it is best to use eightpenny steel-cut flooring nails; for $^{1}/_{2}$-inch flooring, sixpenny nails should be used. Other types of nails have been developed in recent years for nailing of flooring, among these being the annularly grooved and spirally grooved nails. In using these nails, it is a good idea to check with the floor manufacturer's recommendation as to size and diameter for a specific use (Figure 2-3). The nail is driven straight down through the board at the groove edge. The nails should be driven into the joist and near enough to the edge so that they will be covered by the base or shoe molding. The first strip of flooring can be nailed through the tongue.

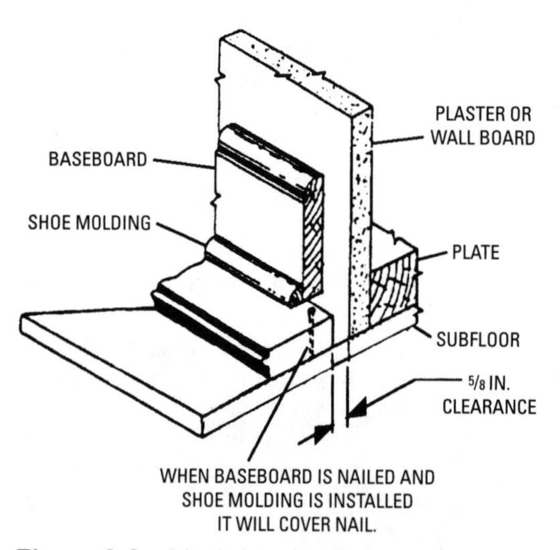

Figure 2-3 Method of laying the first strips of wood flooring.

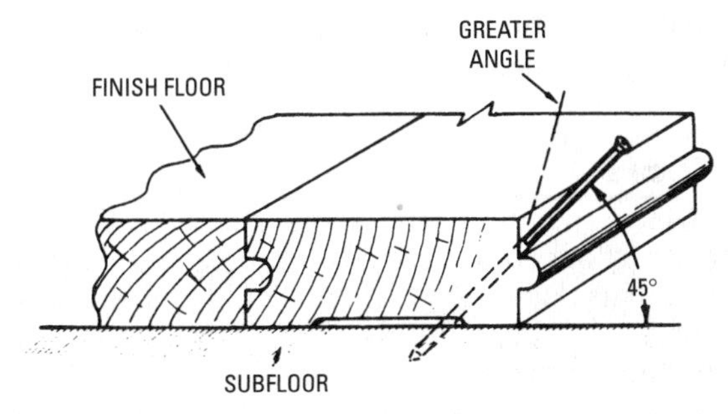

Figure 2-4 Nailing method for strip wood flooring.

Figure 2-4 shows a nail driven in at an angle of between 45° and 50° where the tongue adjoins the shoulder. Do not try to drive the nail down with a hammer, as the wood may be easily struck and damaged. Instead, use a nail set to finish off the driving. Figure 2-5 shows the position of the nail set commonly used for the final driving. To avoid splitting the wood, it is sometimes necessary to pre-drill the holes through the tongue. This will also help to drive the nail easily into the joist. For the second course of flooring, select a piece so that the butt joints will be well-separated from those in the first course. For floors to be covered with rugs, the long lengths could be used at the sides of the room and the short lengths in the center where they will be covered.

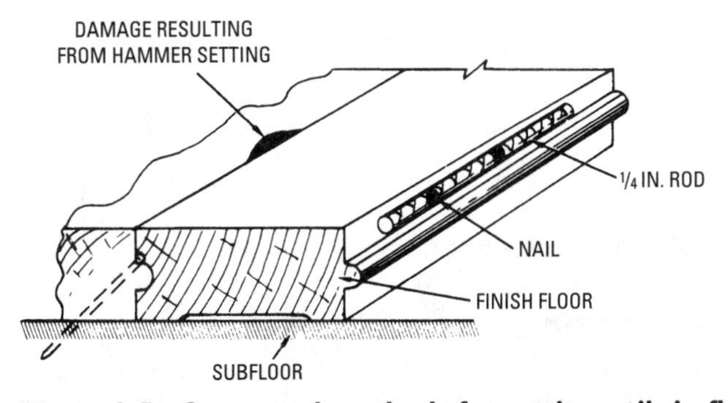

Figure 2-5 Suggested methods for setting nails in flooring.

Each board should be driven up tightly, but do not strike the tongue with the hammer, because this may crush the wood. Use a piece of scrap flooring for a driving block. Crooked pieces may require wedging to force them into alignment. This is necessary so that the last piece of flooring will be parallel to the baseboard. If the room is not square, it may be necessary to start the alignment at an early stage.

Soundproof Floors

One of the most effective sound-resistant floors is called a *floating* floor. The upper or finish floor is constructed on 2 × 2 joists actually floating on glass wool mats (Figure 2-6). There should be absolutely no mechanical connection through the glass wool mat, not even a nail to either the subfloor or the wall.

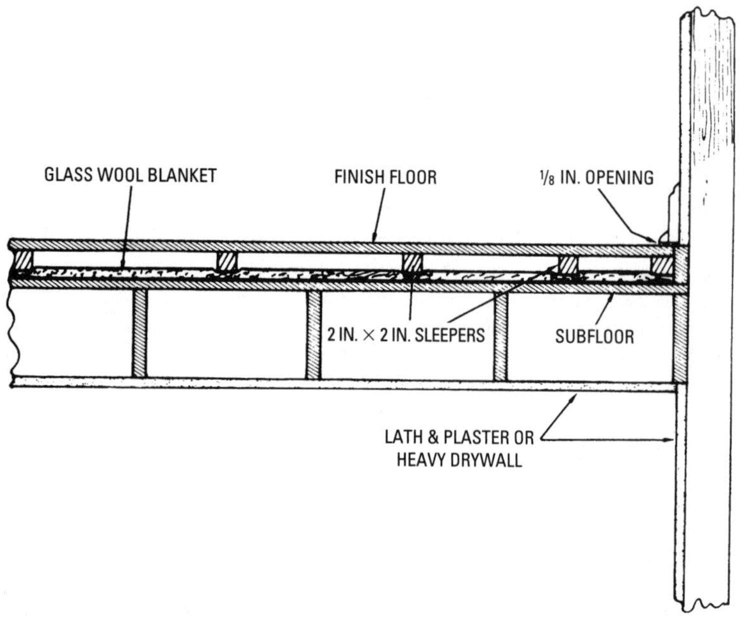

Figure 2-6 A sound-resistant floor.

Wood-Tile Flooring

Flooring manufacturers have developed a wide variety of special patterns of flooring, including parquet, which is nailed in place or has an adhesive backing (Figure. 2-7), that can be used over wood subfloors or concrete slabs. One common type of floor tile is a block 9 inches square and $^{11}/_{32}$ inch thick that is made up of several individual strips of flooring held together with glue and splines. Two

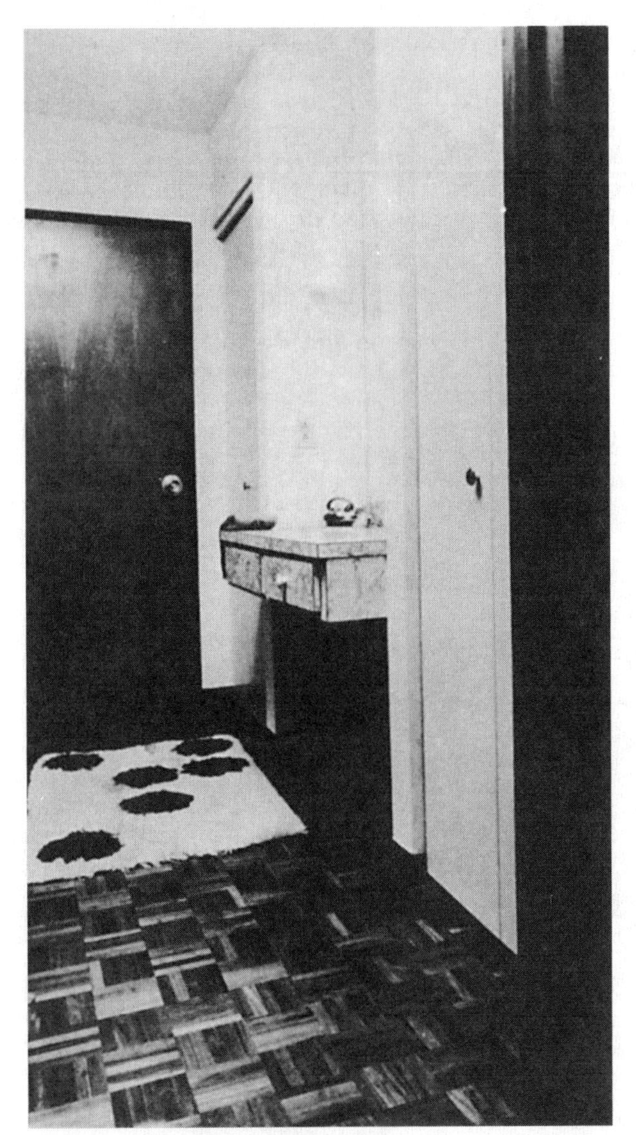

Figure 2-7 Wood-tile flooring. (Courtesy of United Gilsonite)

edges have a tongue and the opposite edges are grooved. Numerous other sizes and thicknesses are available. In laying the floor, the direction of the blocks is alternated to create a checkerboard effect. The manufacturer supplies instructions for laying its tile. Follow them carefully. When the tiles are used over a concrete slab, a vapor barrier should be used. The slab should be level, and should be thoroughly cured and dry before the wood tile is laid.

Ceramic Tile

Ceramic tile (Figure 2-8) is made in different colors and with both glazed and unglazed surfaces. It is used as a covering for floors in

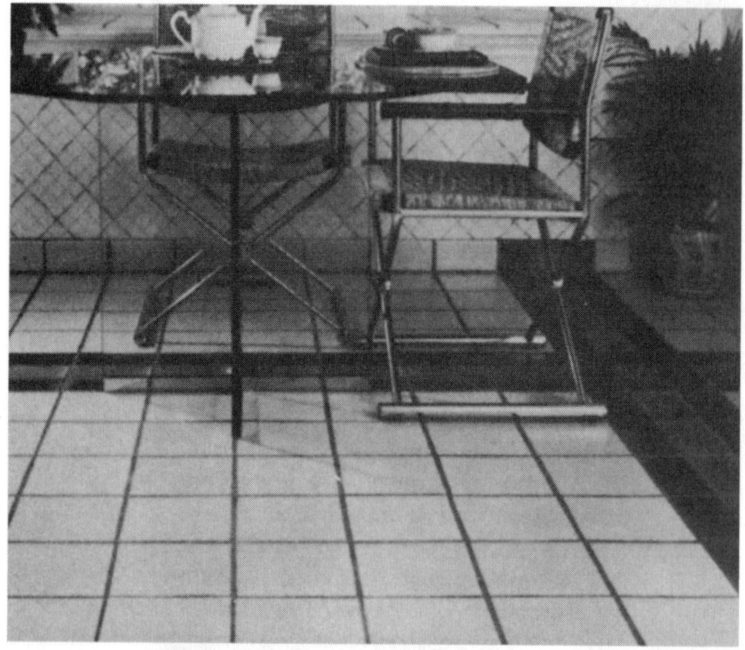

Figure 2-8 Ceramic tile. (Courtesy of Olean Tile)

bathrooms, entryways, kitchens, and fireplace hearths. Ceramic tile presents a hard and impervious surface. When ceramic-tile floors are used with wood-frame construction, a concrete bed of adequate thickness must be installed to receive the finishing layer.

Installation of tile is done with adhesive. Check with your dealer for specific instructions.

Other Finished Floorings

Resilient flooring is so-called because it gives when you step on it. There are 12-inch tiles available for installing with adhesive, as well as adhesive-backed tiles. Today, most tile is vinyl and comes in a tremendous variety of styles, colors, and patterns.

Resilient flooring also comes in sheets or rolls 12 feet wide. It, too, is chiefly vinyl and comes in a great array of styles and colors. It is more difficult to install than tile (Figs. 2-9 to 2-13).

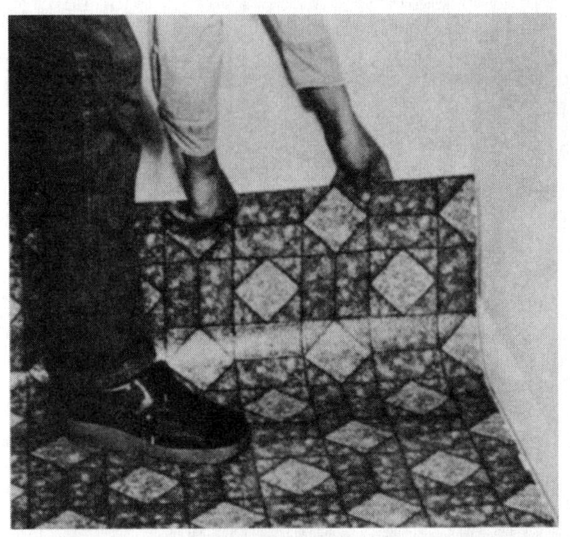

Figure 2-9 Resilient flooring is first laid in a room.

Resilient sheet flooring comes in several qualities, and you should compare competing materials before you buy. Vinyl resilient flooring may be installed anywhere in the house, above or below grade, with no worry about moisture problems. An adequate subfloor (usually of particleboard or plywood) is required.

Also available is carpeting. Wall-to-wall carpet installation is usually contracted out. Carpeting is available for indoor and outdoor use.

Other flooring materials include paint-on coatings and paints. Paint is normally used in areas where economy is most important.

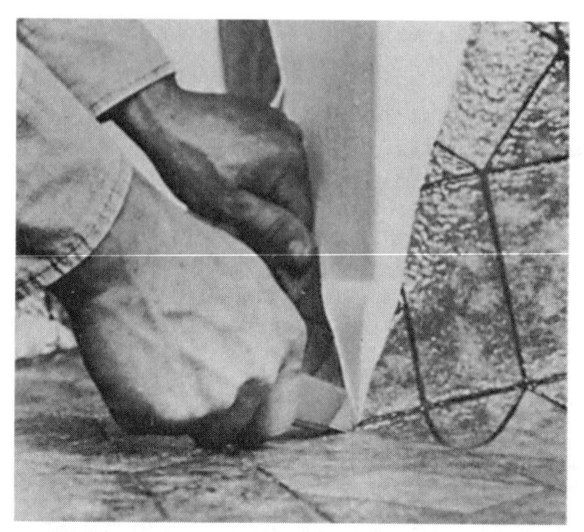

Figure 2-10 Relief cuts are made where flooring abuts walls. (Courtesy of Congoleum)

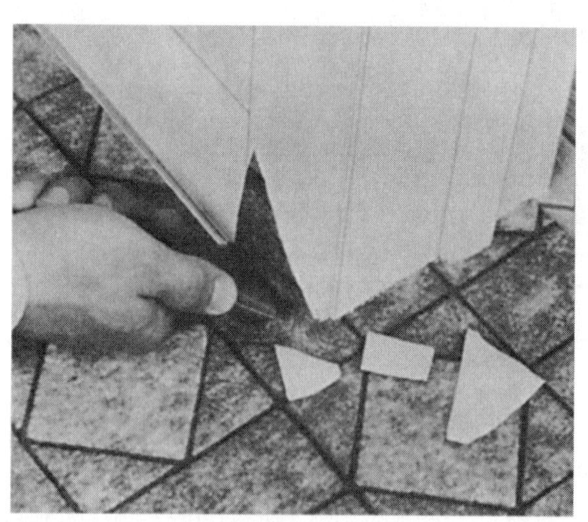

Figure 2-11 A utility knife is used to trim flooring. (Courtesy of Congoleum)

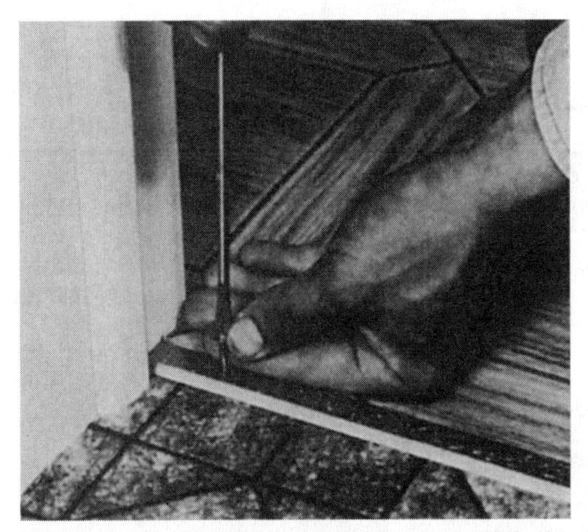

Figure 2-12 Metal threshold is screwed in place. (Courtesy of Congoleum)

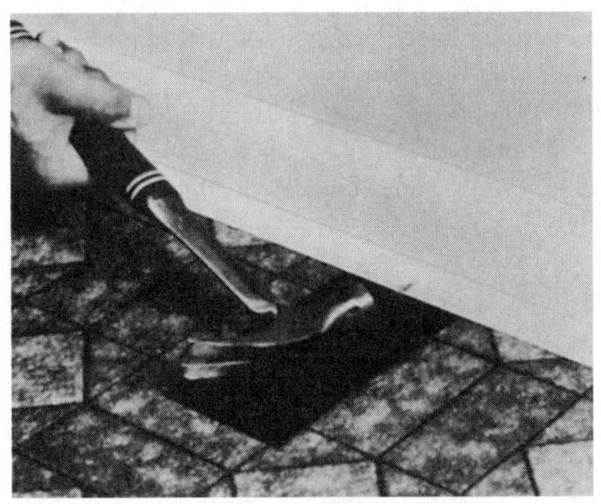

Figure 2-13 The final step is to reinstall molding. (Courtesy of Congoleum)

Summary

Hardwood floorings most commonly used are red and white oak, hard maple, beech, and birch. Most flooring is $^{25}/_{32}$-inch thick with a face width of $2^1/_4$ inches. The finished flooring is usually installed after all plastering is complete, after windows and doors are in place, and after most of the interior trim has been installed.

When the flooring is ready for installation, the subfloor should be thoroughly cleaned. A layer of deadening felt or building paper will deaden the sound, plus it will help to eliminate floor squeaks. Where a crawl space is used, it will increase the warmth of the floor by preventing air infiltration.

Carpeting is usually contracted out. Resilient flooring is available in many types and patterns and finishes from wax to no-wax. Paint is used as a floor covering only where economy is most important.

Review Questions

1. What type of wood is generally used for finish flooring?
2. What is the thickness of hardwood flooring?
3. Why is building paper or deadening felt used between flooring?
4. How are soundproof floors constructed?
5. Hardwoods and softwoods are available as strip flooring in a variety of widths and thicknesses, as well as random-width planks, ______, and block flooring.
6. When should the flooring be laid?
7. How is strip flooring laid?
8. What is the purpose of felt or heavy building paper in the laying of a floor?
9. What is meant by a 2 × 2 sleeper?
10. How is strip wood flooring nailed to the subfloor?

Chapter 3

Walls and Ceilings

The most common material for building interior walls and ceilings is drywall, also known as the following:

- Gypsum board
- Plasterboard
- Wallboard
- Sheetrock (a brand name)

It is installed with much less time, effort, and expense than plaster.

Plaster is still used in high-end work. However, it takes a good deal of skill to apply plaster correctly, and it is time and labor consuming. On the other hand, it is generally conceded that a plaster wall is better than plasterboard in sound transmission and smoothness.

Plaster

When plaster is used to form the wall surface, it is held in place by lath. Two kinds of lath are in common use:

- Expanded metal
- Gypsum

Plaster Reinforcing

Since some drying may take place in wood framing members after the house is completed, some shrinkage can be expected. This, in turn, may cause the plaster to crack around openings and corners. To minimize this cracking, expanded metal lath is used in certain key positions over the plaster base material as reinforcement. Strips of expanded metal lath may be used over the window and door openings (Figure 3-1). A strip 8 inches × 20 inches placed diagonally across each upper corner and nailed lightly in place should be effective (Figure 3-2).

Inside corners at the juncture of walls and ceilings should be reinforced with corners of metal lath or wire fabric (Figure 3-3), except where special clip systems are used for installing the lath. The minimum width of the lath in the corners should be 5 inches, or $2^1/_2$ inches on each surface or internal angle, and should be lightly nailed in place. Corner beads (Figure 3-4) of expanded metal lath or perforated metal should be installed on all exterior corners. They should be applied plumb and level. The bead acts as a leveling edge

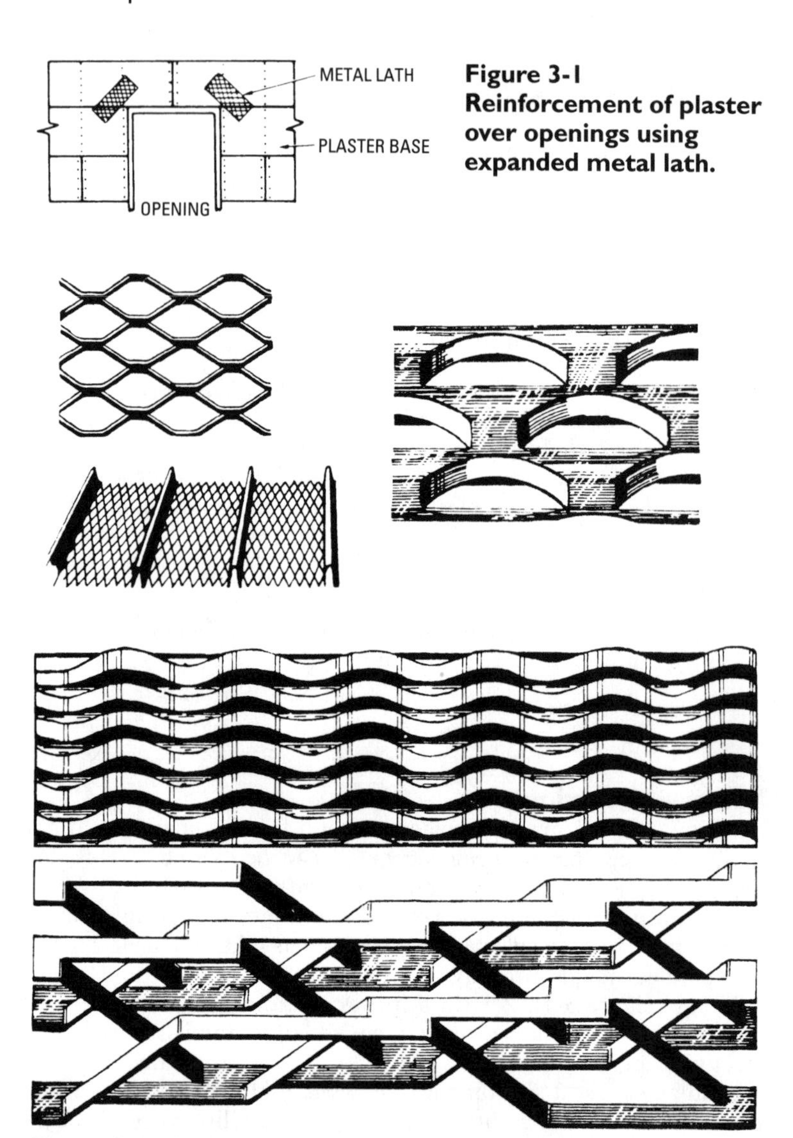

Figure 3-1 Reinforcement of plaster over openings using expanded metal lath.

Figure 3-2 Various types of expanded metal lath.

when the walls are plastered and reinforces the corner against mechanical damage.

Metal lath should be used under large flush beams (Figure 3-5). It should extend well beyond the edges. Where reinforcing is required over solid wood surfaces (such as drop beams), the metal lath should

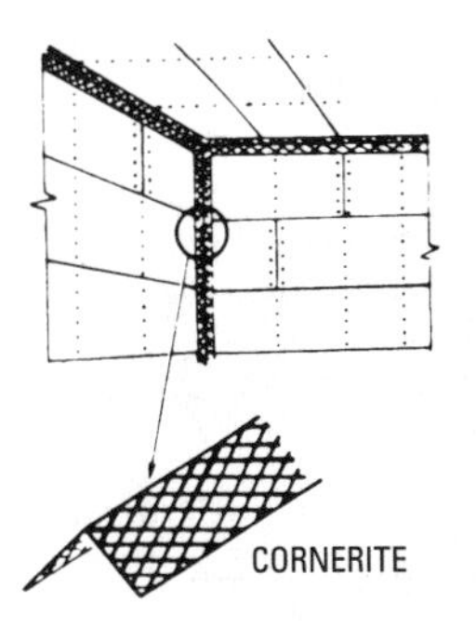

Figure 3-3 Use of cornerites made from metal lath.

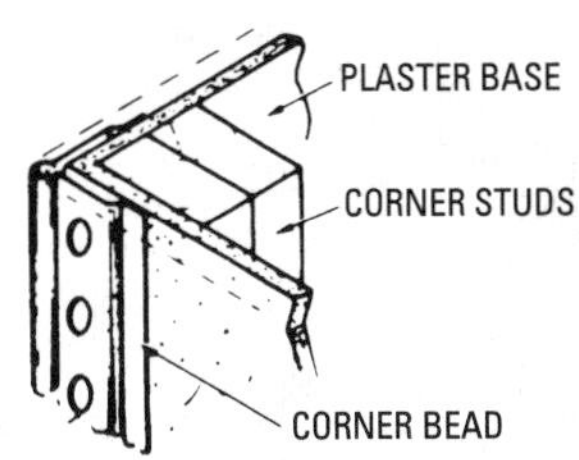

Figure 3-4 Use of a corner bead.

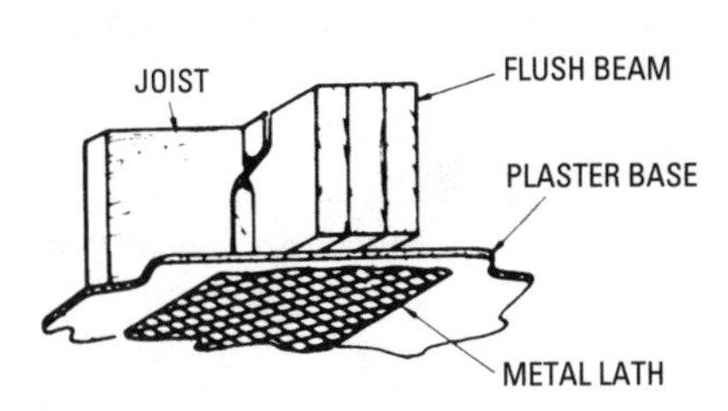

Figure 3-5 Use of expanded metal lath under large flush beams.

either be installed on strips, or else self-furring nails should be used to set the lath out from the beam. The lath should be lapped on all adjoining gypsum lath surfaces.

Expanded metal lath applied around a bathtub recess for ceramic tile application should be used over a paper backing (Figure 3-6). Studs should be covered with 15-pound asphalt saturated felt applied shingle style. The scratch coat should be Portland-cement plaster, of $^{5}/_{8}$-inch minimum thickness, and integrally waterproofed. The scratch coat should be dry before the ceramic tile is applied. Expanded metal lath consists of sheet metal that has been slit and expanded to form innumerable openings for the keying of the plaster. It should be painted or galvanized, and its minimum weights for 16-inch stud or joist spacing are as shown in Table 3-1.

Metal lath is usually 27 inches × 96 inches in size. Other materials (such as wood lath, woven wire fabric, and galvanized wire fabric)

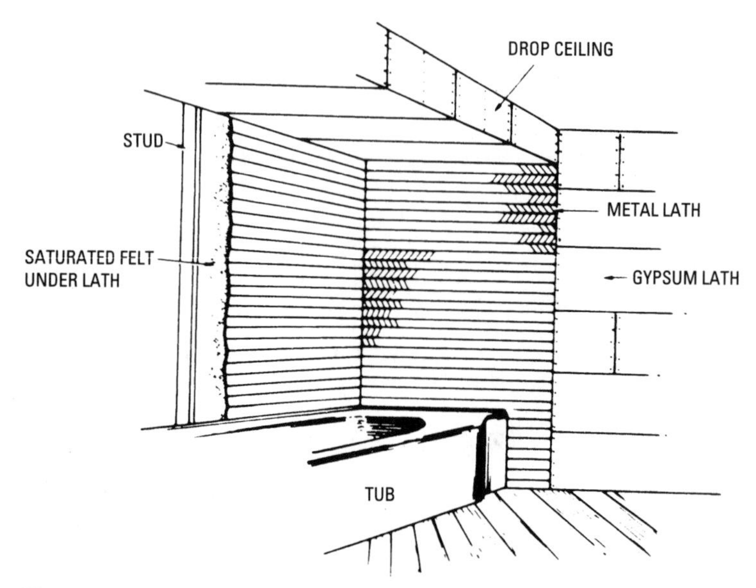

Figure 3-6 Application of expanded metal lath around bathtub for the installation of ceramic tile.

Table 3-1 Minimum Weights

Use	*Pounds per Square Yard*
Walls	2.5
Ceilings	3.4
Ceiling with flat rib	2.75

may also be used as a plaster base. Figure 3-7 shows various types of woven wire lath.

Installing Gypsum Base

A plaster finish requires a base on which the plaster can be spread. The base must have bonding qualities so the plaster will adhere to (or will be keyed to) the base. That means the base has been fastened to the framing members. One of the popular types of plaster base that may be used on sidewalls and ceilings is gypsum-board lath. Such lath is generally 16 inches × 48 inches and is applied horizontally (across) the frame members. This type of board has a paper face with gypsum filler. For studs and joists with spacing of 16 inches on center, $^3/_8$-inch thickness is used, and for 24 inches on center spacing,

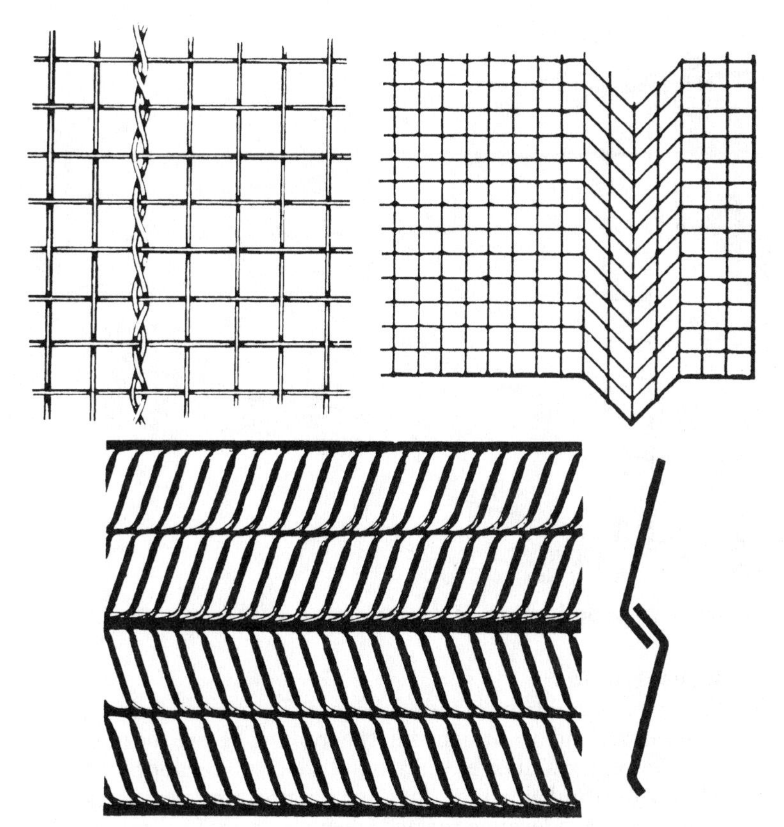

Figure 3-7 Various types of woven wire lath.

$^{1}/_{2}$-inch thickness is used. This material can be obtained with a foil backing that serves as a vapor barrier, and if it faces an air space, it has some insulating value. It is also available with perforations that improve the bonding strength of the plaster base.

Gypsum lath should be applied horizontally, with the joints broken (Figure 3-8). Vertical joints should be made over the center of the studs or joists, and should be nailed with 13-gage gypsum lathing nails $1^{1}/_{8}$ inches long and having a $^{3}/_{8}$-inch flat head. Nails should be spaced 4 inches on center and should be nailed at each stud or joist crossing. Lath joints over the heads of doors and window openings should not occur at the jamb lines. Insulating lath should be installed much like gypsum lath, except that 13-gage $1^{1}/_{4}$-inch blued nails should be used.

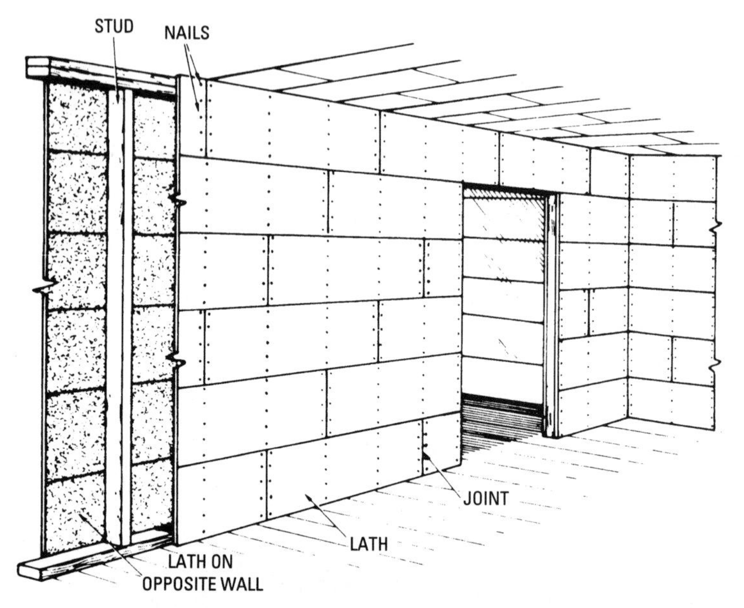

Figure 3-8 Application of gypsum plaster base.

Plaster Grounds

Plaster grounds are strips of wood that are the same thickness as the lath and plaster and are attached to the framing before the plaster is applied. Plaster grounds are used around window and door openings as a plaster stop and along the floor line for attaching the baseboard. They also serve as a leveling surface when plastering and as a nailing base for the finish trim (Figure 3-9).

There are two types of plaster grounds: those that remain in place (Figure 3-10) and those that are removed after plastering is completed (Figure 3-11). The grounds that remain in place are usually $^7/_8$-inch thick and may vary in width from 1 inch around openings to 2 inches for those used along the floor line. These grounds are nailed securely in place before the plaster base is installed.

Nailing Lath

Lath nailers are horizontal or vertical members to which lath, gypsum boards, or other covering materials are nailed. These members are required at interior corners of the walls and at the juncture of the wall and ceiling. Vertical lath nailers may be composed of studs so arranged as to provide nailing surfaces (Figure 3-12). This construction also provides a good tie between walls.

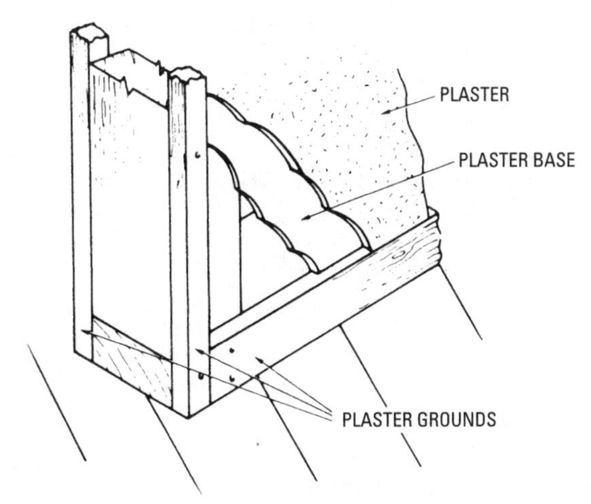

Figure 3-9 Plaster grounds used at doors and floor line.

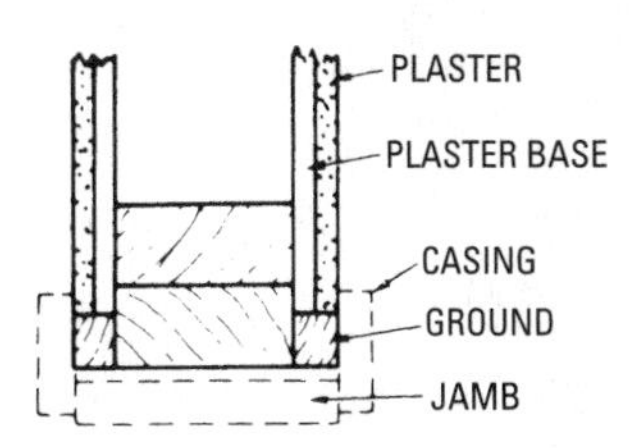

Figure 3-10 Plaster grounds that stay in place and are used as a nailing base for the trim.

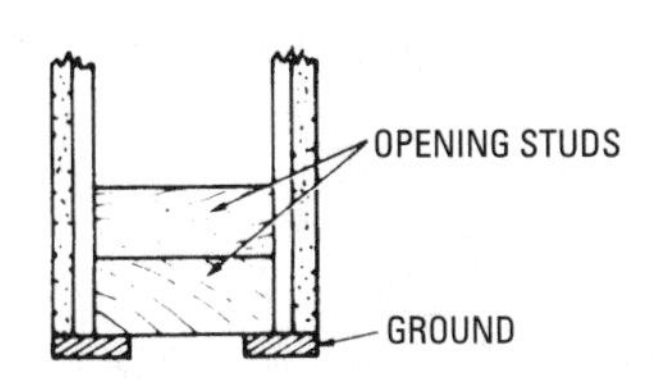

Figure 3-11 Plaster grounds that are removed after plastering.

Another vertical nailer construction consists of a 2-inch × 6-inch lathing board that is nailed to the stud of the intersecting wall (Figure 3-13). Lathing headers are used to back up the board. The header should be toenailed to the stud. Doubling of the ceiling joist over the wall plates provides a nailing surface for interior finish material, as shown in Figure 3-14. Walls may be tied to the ceiling framing in this method by toenailing through the joists into the wall plates.

Another method of providing nail surface at the ceiling line is similar to that used on the walls. A 1-inch × 6-inch lathing board is

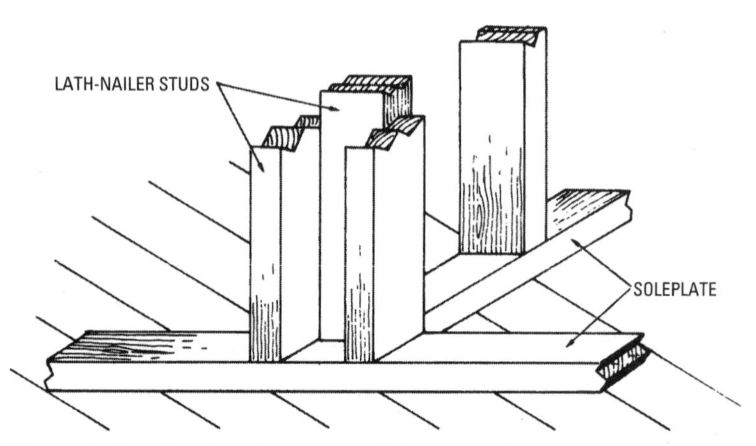

Figure 3-12 Lath nailers at wall intersections.

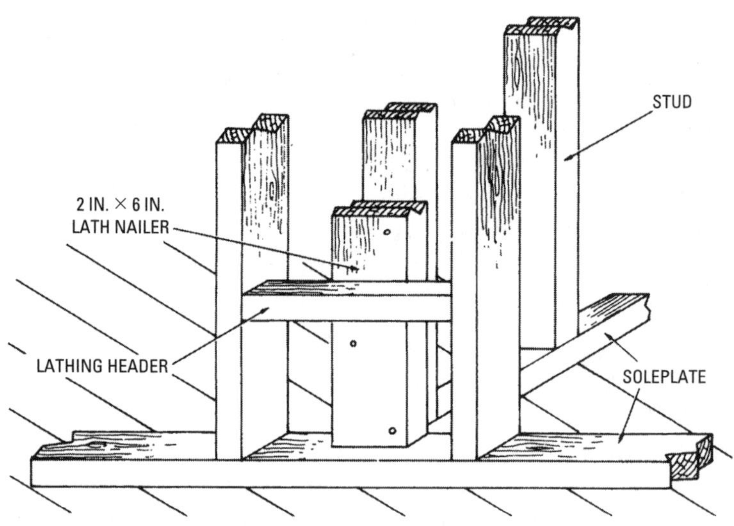

Figure 3-13 Another method of installing lath nailers at intersection walls.

nailed to the wall plate, as shown in Figure 3-15. Headers are used to back up this board, and the header, in turn, can be tied to the wall by toenailing into the wall plate.

Plastering Material and Method of Application

Plaster for interior finishing is made from combinations of sand, lime or prepared plaster, and water. Waterproof-finish wall materials are

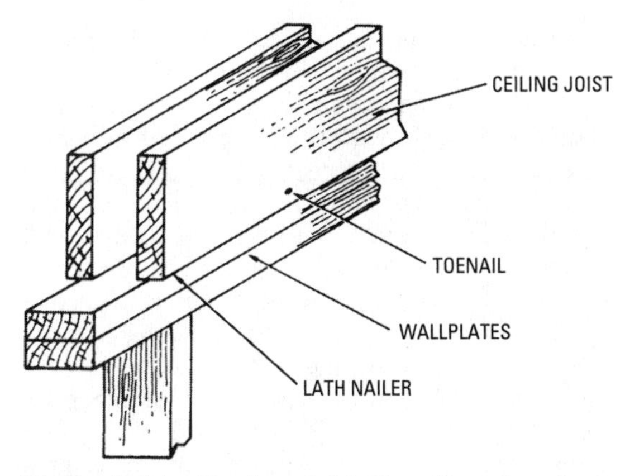

Figure 3-14 Horizontal lath nailers for plaster base formed by ceiling joists.

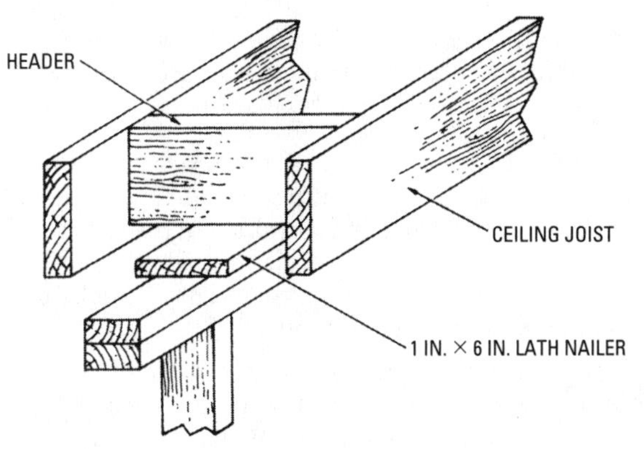

Figure 3-15 Horizontal lath nailers for ceiling provided by lathing boards.

available and should be used in bathrooms, especially in showers or tub recesses when tile is not used, and sometimes in the kitchen wainscot. Plaster should be applied in three coats, or a two-coat double-up work. The minimum thickness over a lath or masonry should be $^1/_2$ inch. The first plaster coat over the metal lath is called the *scratch coat*, and is scratched after a slight set has occurred to ensure a good bond for the second coat. The second coat is called the *brown* or *leveling coat*, and it is during the application of this coat that the leveling is done.

The double-up work, combining the scratch and brown coat, is used on gypsum or insulating lath. The leveling and plumbing of walls and ceiling are done during the application of this work. The final or finish coat consists of two general types: the sand-float and the putty finish. In the *sand-float finish*, lime is mixed with sand and results in a textured finish, with the texture depending on the coarseness of the sand used. *Putty finish* is used without sand and has a smooth finish. This is commonly used in kitchens and bathrooms, where a gloss paint or enamel finish is often employed, and in other rooms where a smooth finish is desired. The plastering operation should not be done in freezing weather without the use of constant heat for protection from freezing. In normal construction, the heating unit is in place before plastering is started.

Insulating plaster (consisting of vermiculite, Perlite, or other aggregate used with the plaster mix) may also be used for wall and ceiling finishes. This aggregate properly mixed with the plaster produces small hollow air pockets, which act as insulating material. The vermiculite is a material developed from a mica base that is exploded in size, and, when mixed with the plaster, reduces the added weight in a conventional plastered wall or ceiling.

The following points in plaster maintenance are worthy of attention:

- In a newly constructed house, a few small plaster cracks may develop during or after the first heating season. These cracks are usually caused by the drying and shrinking of the structural members. For this reason, it is advisable to wait until after a heating season before painting the plaster. These cracks can then be filled before painting has begun.
- Because of the curing period ordinarily required for a plastered wall, it is not advisable to apply oil-base paints until at least 60 days after plastering is completed. Water-mix (or resin-base) paints may be applied without the necessity of an aging period.
- Large plaster cracks often indicate a structural weakness in the framing. One of the common areas that may need correction is around a basement staircase. Framing may not be adequate for the loads of the walls and ceilings. In such cases, the use of an additional post and pedestal may be required to correct this fault. Inadequate framing around fireplaces, chimney openings, and joists that are not doubled under partitions are other common sources of weakness.

In addition to the plastering methods described, there has been one other development in plastering that reduces the plastering process to "one coat" and that gives the process its name. Here, "green" or "blue" $^5/_8$-inch gypsum board is installed like regular gypsum board, and the premixed plaster material (which comes in 5-gallon cans) is applied by the plasterer. That's it. The applied material cures very hard, and the gypsum board has the tooth to keep it in place.

Drywall

The use of drywall board requires that the studs and ceiling joists be in alignment. Good solid rigid sheathing on the exterior walls will accomplish this on the studs, and strong back bracing on the ceilings joists (Figure 3-16) will keep the ceiling joists level and in alignment.

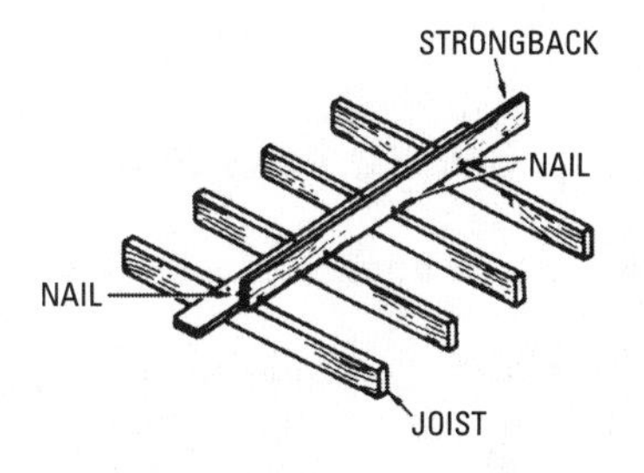

Figure 3-16 Strong back bracing for ceiling joists.

Gypsum board (or *drywall*) is a sheet material composed of gypsum filler faced with paper. Sheets are usually 4 feet wide and can be obtained in lengths of from 8 to 12 feet. The edges along the length of the sheet are recessed to receive joint cement and tape. Although gypsum board may be used in $^3/_8$-inch thickness, a $^1/_2$-inch board offers greater resistance to deflection between framing members.

Gypsum board may be applied to the walls either vertically or horizontally.

With the *vertical method,* applications with 4-foot-wide sheets will cover three studs when they are spaced 16 inches on center. The edges should be centered on the studs, and only moderate contact should be made between the edges of the sheet. Large-head, blued $1^5/_8$-inch drywall nails should be used with $^1/_2$-inch gypsum board, and $1^3/_8$-inch drywall nails should be used with $^3/_8$-inch board. Nails should be spaced 6 to 8 inches for sidewalls, and 5 to 7 inches for ceiling application (Figure 3-17).

The *horizontal method* of application is best adapted to rooms in which full-length sheets can be used, to minimize the number of vertical joints. When joints are necessary, they should be made at windows and doors. Nail spacing is the same as that used in vertical application. Where framing members are more than 16 inches on

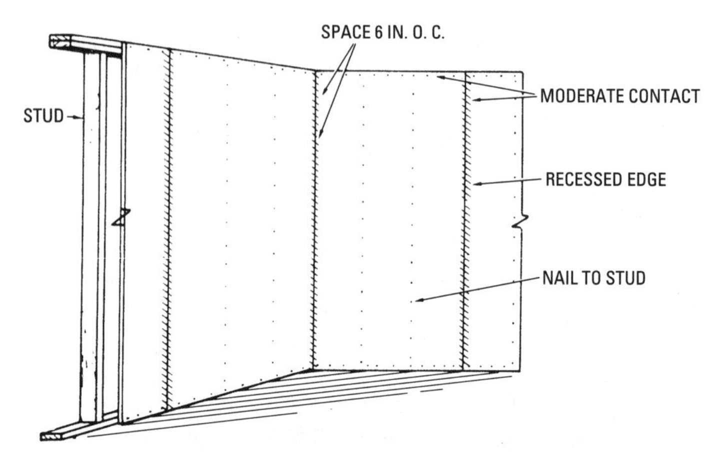

Figure 3-17 Application of vertically applied drywall finish.

center, solid blocking should be used along the horizontal joint (Figure 3-18).

Another method of gypsum board application includes an undercourse of 3/8-inch material applied vertically by nailing. The finish 3/8-inch sheet is applied horizontally, usually in room-size lengths

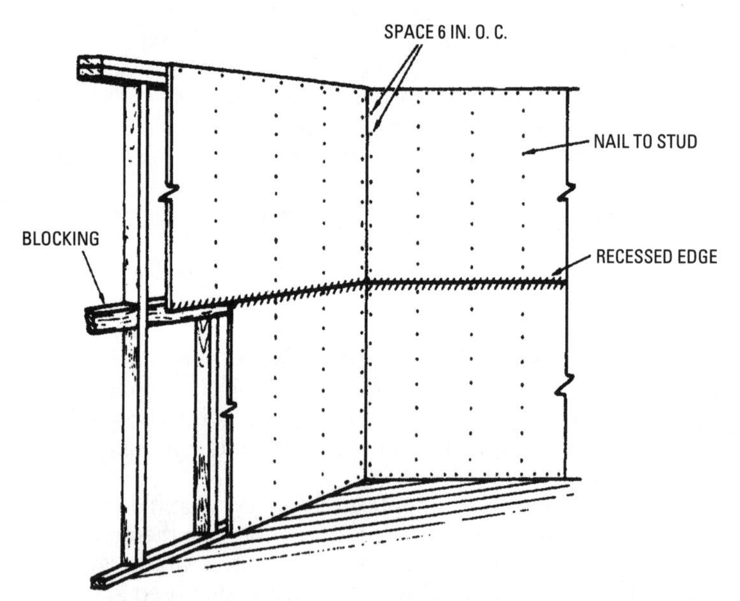

Figure 3-18 Application of horizontally applied drywall finish.

by means of an adhesive, using only enough nails to hold the sheet in place until the adhesive is dry. The manufacturer's recommendations should be followed. In the finishing operation, the nails should be set about $^{1}/_{16}$ inch below the face of the board. This set can be obtained by using a slightly crowned hammer (Figure 3-19). Setting the nail in this manner, a slight dimple is formed in the face of the board without breaking the paper surface. The setting of the nail is particularly important for center nails, since the edge nailing will be covered with tape and joint cement.

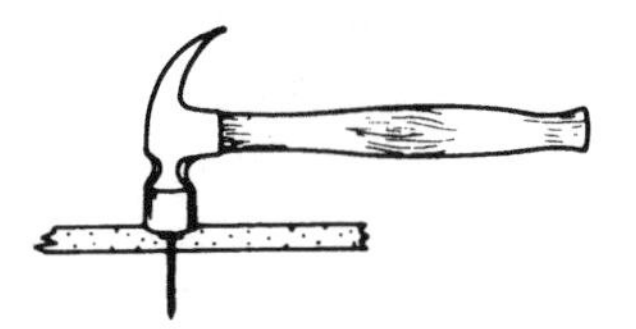

Figure 3-19 Nail set with crowned hammer on drywall.

Joint cement is available in powder form to be mixed with water to a soft putty consistency, as well as a prepared vinyl mix. The procedure for taping joints (Figure 3-20) is as follows:

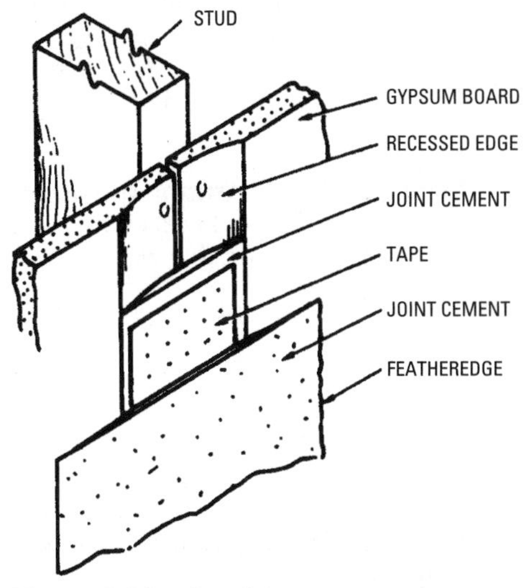

Figure 3-20 Applying tape and cement to wall joints.

1. Use a wide, broad knife (4 inches to 6 inches) and spread the cement in the recess starting at the top of the wall.
2. Press the tape into the recess with the knife until the joint cement is forced through the perforations.
3. Cover the tape with additional cement, feathering the outer edge.

4. Allow the cement to dry and apply a second coat and feather the edges. A steel trowel is sometimes used. For best results, a third coat may be applied, feathering beyond the second coat.
5. After the joint cement has dried, sand it smooth.
6. Hide nail indentations in the center of the board, fill with joint cement, and sand smooth when dry.

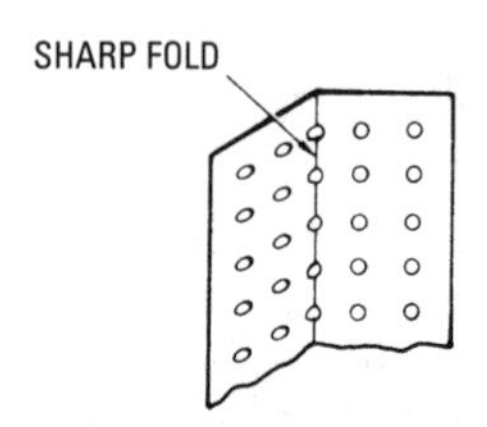

Figure 3-21 Folding joint tape for use on interior corners.

Interior corners may be treated with tape. Fold the tape down the center to a right angle (Figure 3-21). Apply cement to both sides and press the perforated tape into the cement. Smooth the tape down with a trowel. There is a special corner trowel for this. After the cement has dried, apply a second coat. Sand the cement when dry, and apply a third coat if needed. The interior corners between the wall and ceiling may be concealed by the use of a molding. When molding is used (Figure 3-22), it is not necessary to tape the joints. Metal corner beads should be used on exterior corners and for openings where a casing or trim is not used.

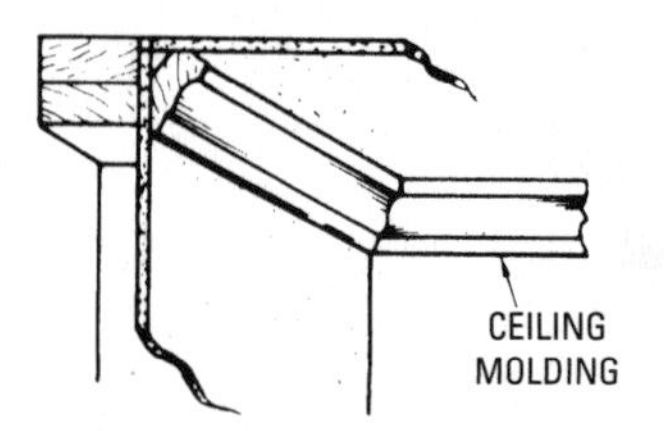

Figure 3-22 Illustrating the use of molding at the wall and ceiling corner when using drywall construction.

Masonry Walls as Plaster Base

Masonry provides an excellent base for plaster. The surface of any masonry unit should be rough to provide a good mechanical key and should be free from paint, oil, dust, dirt, or any other material that might prevent a satisfactory bond. Proper application of plaster requires that

- The plaster base material bonds and becomes an integral part of the base to which it is applied
- It be used as a thin reinforced base for the finished surface

Old masonry walls that have been softened by weathering or surfaces that cannot be cleaned thoroughly must be covered with metal reinforcement before applying the plaster. Metal reinforcement should be applied to wood furring strips (Figure 3-23). The metal reinforcement should be well-braced and rigid to prevent cracking the plaster. This type of construction will give some insulating qualities because of the air space between masonry and plaster.

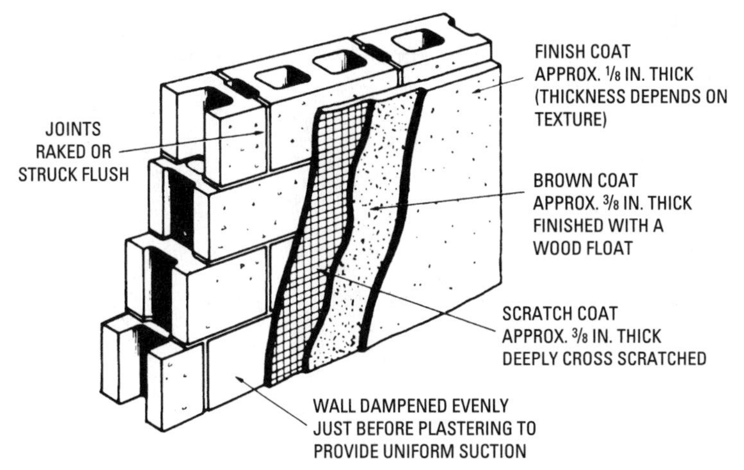

Figure 3-23 The application of plaster to masonry.

Summary

Metal lath is generally used around openings and corners to minimize cracking the plaster. Since wood draws moisture from the plaster, metal lath is used to reinforce the plaster. This type of material is also used to cover wide cracks and open joints in lath.

Gypsum board is another type of lath, which is usually 16 inches × 48 inches, and is applied horizontally to the wall studs. It is obtained in 3/8-inch and 1/2-inch thicknesses, depending on wall stud spacing.

Drywall is the most popular finish material. It is obtained in 4-foot × 8-foot or 4-foot × 12-foot sheets. The edges along the length of the sheet are recessed to receive the joint tape and cement.

Review Questions

1. What are plaster grounds and why are they used?
2. Why is gypsum board or plasterboard used instead of wood lath?
3. What is drywall and why is it used?

4. Why is metal lath used and where is it installed?
5. What is joint tape and how is it used?
6. Where is plaster used today?
7. What thicknesses does drywall usually come in?
8. What is a cornerite?
9. A plaster finish requires a base on which the ____ can be spread.
10. Should gypsum lath be applied horizontally or vertically?

Chapter 4

Kitchen Cabinets

The kitchen as a workspace must be efficient. As an entertainment center, it must be attractive. As an investment, it must increase the home's value. For many families, it must be versatile, function as a social, communications center, and hobby area, as well as serve the dual purpose of a cooking and dining area.

Kitchen Design

The design should incorporate convenience and ample storage. Safety should be a factor, as well as the attractiveness of the kitchen, because it's probably the most important room in the house. More elements (cabinets, appliances, fixtures, plumbing, wiring, and furniture) go into the kitchen than any other room in the home. Chances are more working hours are spent there than anywhere else. Creating a new kitchen in a new or existing home is a major task that requires sound planning. Careful planning can eliminate costly mistakes.

Evaluating Needs

Creating a new kitchen may seem like an overwhelming process, but it needn't be if you take it one step at a time. Begin by evaluating the needs of the users.

Who will use the kitchen? Is it a setting for family gatherings or the private domain of a gourmet chef? Will it still be workable when there are two or more cooks rather than just one.

In doing the evaluation, cooking needs should be the starting point. How often the cooking is done will affect the size, layout, and type of equipment.

For example, if the family will enjoy sharing meal preparation, include two sinks and built-in cutting and chopping boards strategically scattered throughout the kitchen to maximize food-preparation areas.

Entertainment needs also are important. At parties, guests naturally gravitate toward the kitchen. Small intimate gatherings might be easily accommodated by an entertainment bar built right into an island counter or a number of other places in the room.

A preference for large, catered affairs might necessitate installation of two ovens for preparing large quantities of food. Likewise, wide-shelved refrigerators would be helpful in holding large party platters. Storage space for linens also might be a feature to consider.

Will the kitchen be used by a working person without much time for meal preparation? Include a microwave oven and lots of freezer space.

Are children going to be active in the kitchen? If so, easy-to-clean surfaces are necessary. Other features might include a microwave oven for after-school snacks, and a desk or counter where children could do homework or school projects.

Design Elements

Once you have evaluated the needs, the next step is to consider the important work areas: cooking, clean-up, food preparation, and storage. Once these have been identified, arrange them in an efficient, logical pattern.

The Triangle

The fundamental design revolves around a work triangle, formed around the three major kitchen elements—range, refrigerator, and sink (Figure 4-1). More trips are made within this triangle than to any other areas of the kitchen.

For the most efficient layout, kitchen designers recommend that the appliances be placed so that the distance between any two of them is no less than 3 feet and no more than 7 feet, with the total of the triangle sides measuring no more than 22 feet or no less than 12 feet. A greater distance means unnecessary walking; a shorter one means a cramped workspace.

The Cooking Center

The cooking center consists of a range (or cooktop), oven(s), microwave oven, plus the equipment needed to cook and serve food (large ladles, carving knives, serving platters, pots, pans, and spices, for example, as shown in Figure 4-1A). A minimum of 12 inches of counter space on each side of the range or surface unit is necessary, though more space is desirable. In some instances, the oven or microwave may be located outside the work triangle because of its usage. Wherever the separate oven is located, a 15-inch landing counter should be placed to either side for a working space.

The Clean-Up Center

The clean-up center is the area around the sink and the dishwasher (Figure 4-1B). A trash compactor and disposal also would be located here. Dishes, flatware, and glasses are usually stored nearby, so they can go directly from sink or dishwasher to storage. Since food and utensils are cleaned here, many cooks opt for one deep sink bowl for cleaning pots and pans too large to fit in the dishwasher, and a

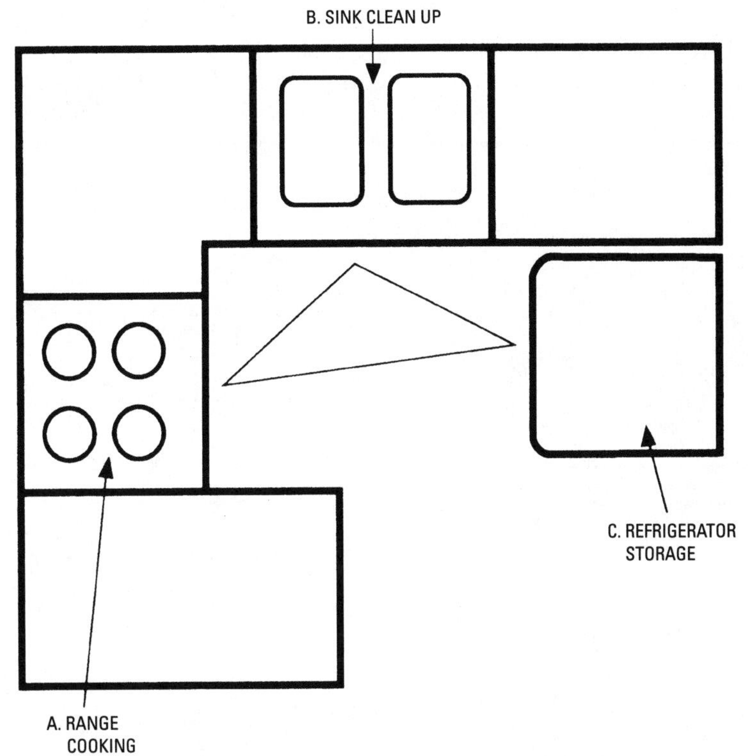

Figure 4-1 Kitchen layout showing the work triangle. *(Courtesy of Kitchen & Bath Source Book)*

second smaller vegetable bowl. Ideally, the clean-up center should be located between the food storage center and the cooking center. Provide at least 18 inches of counter space on one side of the sink and 24 inches of space on the opposite side for this function.

The Refrigeration and Storage Center

The refrigeration and storage center ideally should be located adjacent to the sink, with at least 36 inches of counter space between (Figure 4-1C). This center consists of the refrigerator, freezer, and cabinets for canned, bulk, and packaged foods. Utensils and dishes used in preparing meals should be kept in this area.

While the cooking, clean-up, and refrigeration/storage centers are basic to every kitchen, many kitchens also include a dining area, planning center, and entertainment bar.

Layout

There is a variety of kitchen layouts, each with its own advantages, depending on how you want your kitchen to function (Figure 4-2). Think about the family size, who cooks, what kind of cooking is done, where eating is done, and what architectural constraints exist.

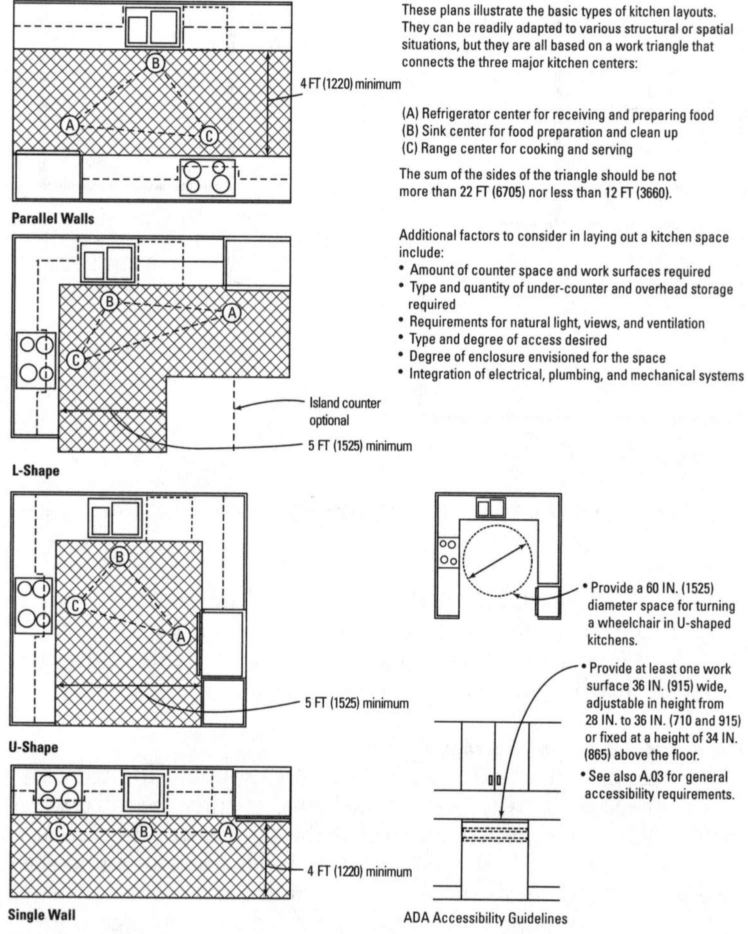

Figure 4-2 Kitchen layouts as prepared by architects. *(Courtesy of Building Construction Illustrated)*

Following are some basic shapes to consider:

- *Peninsulas*—These can be used as dividers between the kitchen and adjacent room areas or to expand any of the kitchen shapes described here (Figure 4-3A).
- *U-shaped kitchen*—This surrounds the cook on three sides with continuous countertop and keeps walk-through traffic out of the work area (Figure 4-3B). This arrangement is a good choice for one person who wants to handle kitchen chores in the most efficient way possible.
- *Island tops*—These are counter-space stretchers (Figure 4-3C). They often are part of a modified L-shaped or U-shaped kitchen. The island can double as clean-up, food preparation, cooking, dining, or entertainment area.
- *L-shaped kitchen*—This gives the cook a generous amount of continuous counter space though generally less than a U-shaped area (Fig 4-4A). With work centers on two adjacent walls, a natural triangle is formed and traffic bypasses the area. This layout is very accommodating for two or more cooks, and often allows space for an island or dining area.
- A *corridor kitchen*—This offers one cook the advantage of an efficient, close grouping of work centers on parallel walls (Figure 4-4B). However, household traffic may cross back and forth through the area, and it may be cramped for two cooks unless you can stretch the width of the aisle between the two working walls.
- A *one-wall kitchen*—This has all work centers stretched along a single wall (Figure 4-4C). Although it is the least efficient kitchen design, it can be used when there are space restrictions.

Cabinet Construction

Today many carpenters find it more economical to have kitchen cabinets custom-built by a cabinet shop or to buy stock cabinets directly from these outlets. Still other carpenters build their own cabinets from scratch.

The size and number of cabinets required depends on the kitchen area available, the amount of cooking to be done, and the distance of the home from store centers (Figure 4-5). Thus, a home that is located within a short distance of stores will not need the storage space required in a home that is remotely located from shopping centers, nor would a small family need the same amount of storage space as that required by a large family.

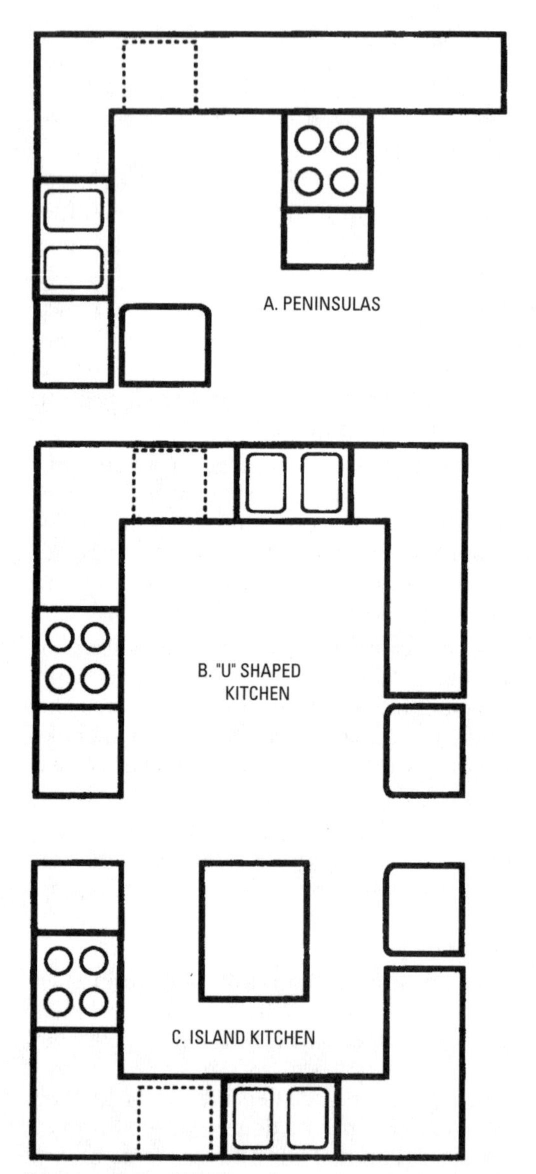

Figure 4-3 Kitchen layouts suitable for larger kitchens. *(Courtesy of Kitchen & Bath Source Book)*

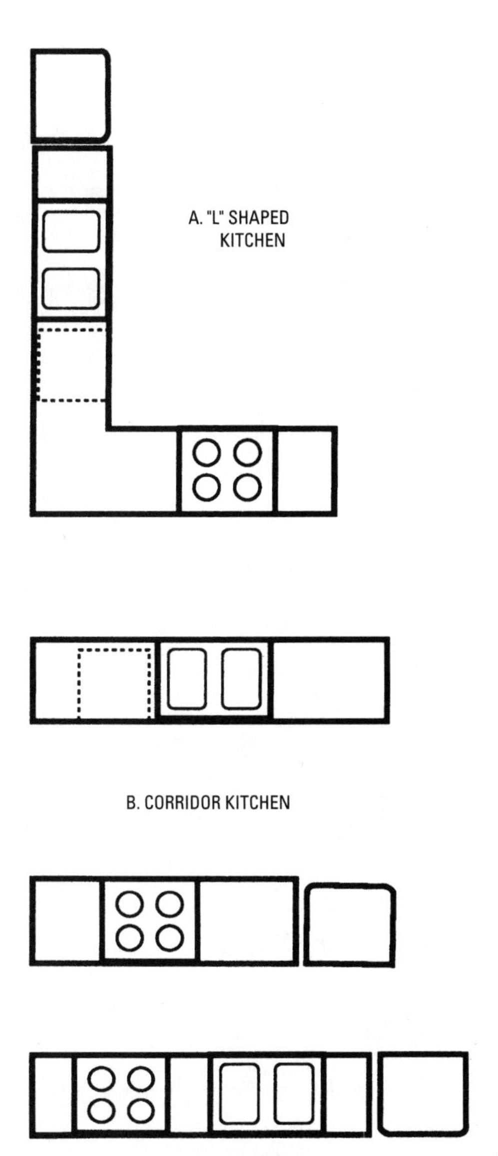

Figure 4-4 Kitchen layouts suitable for smaller kitchens. *(Courtesy of Kitchen & Bath Source Book)*

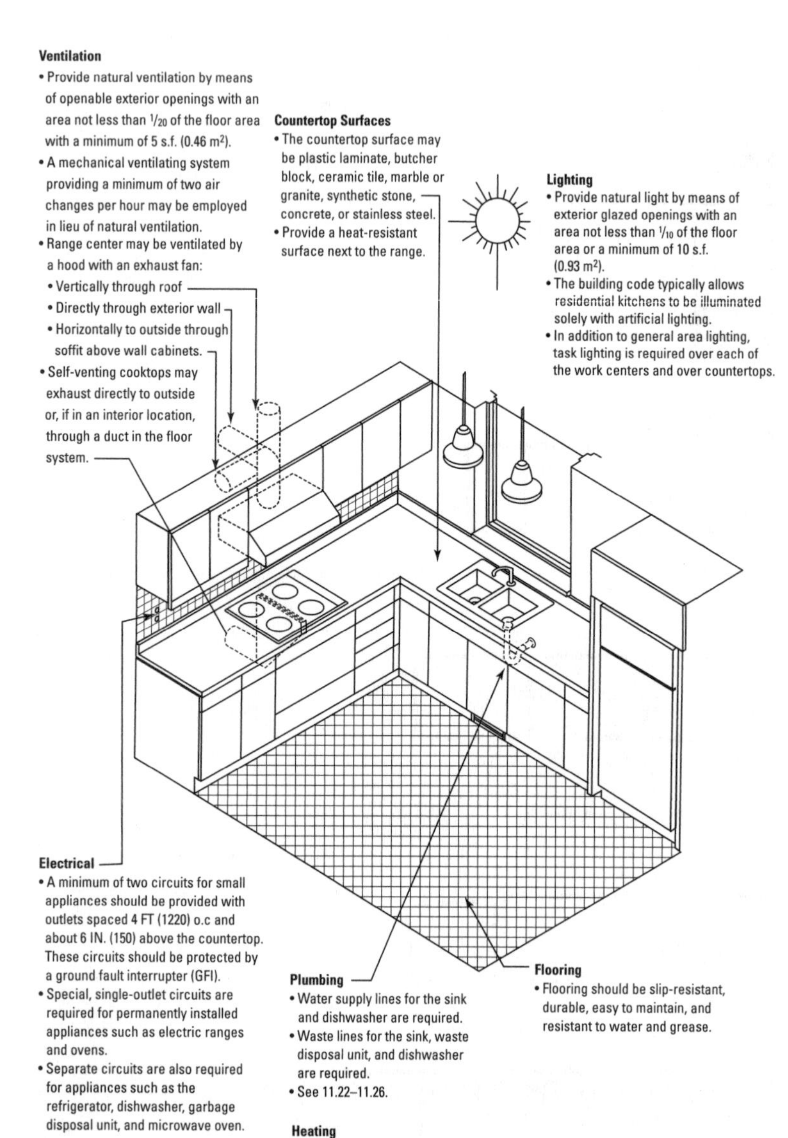

Figure 4-5 The kitchen space. *(Courtesy of Building Construction Illustrated)*

Once the layout has been decided, the next step consists of choosing the number and type of cabinets. The cabinets are usually designed so that they can be made as individual units and installed as such, or several cabinets can be made and installed as a combination. When a combination of cabinets is to be arranged to form a large unit, it is usually better to make a large top of sufficient length to cover the two or more units placed side by side than to make a separate top for each unit.

The number and the location of electrical outlets and fixtures are other important parts of kitchen planning. Wall outlets should be placed above the floor cabinets for easy connection to electrical appliances used in the preparation of food (such as toasters, food mixers, coffeemakers, casseroles, roasters, and broilers). It is a good idea to have several double outlets available to permit the use of as many appliances as possible. Electric and gas ranges commonly come with permanently installed fluorescent lamps that prevent shadow problems when working at the kitchen range. To provide kitchen ventilation and to remove cooking odors, an exhaust fan installed in an outside wall is another useful appliance. Ranges also, of course, come equipped with vents.

Construction Considerations

Although it may readily be admitted that not every carpenter has sufficient skill to design and make artistic cabinets suitable for kitchen use, a great many simple cabinets can be easily made. In general, it is best to start out with a cabinet of simple construction, and, when greater proficiency has been attained in the handling of tools and the working of wood, the more intricate and difficult projects may be undertaken.

Kitchen Wall Cabinets

As the name suggests, kitchen wall cabinets hang on the wall above the floor cabinets and are suitable companion pieces to the floor units. They are simple in construction and may easily be made with simple tools (Figure 4-6). The illustrated design permits construction in widths ranging from 16 to 36 inches, depending on the available and desired space.

For the most convenient usage, it has been found that cabinets of this type should be proportioned so that the top of the cabinet will be approximately 7 feet above the floor surface. Cabinets ranging from 16 to 24 inches in width usually require only one door, while cabinets of 28 to 36 inches in width require two doors. It should be noted, however, that regardless of size, all cabinets are similar in

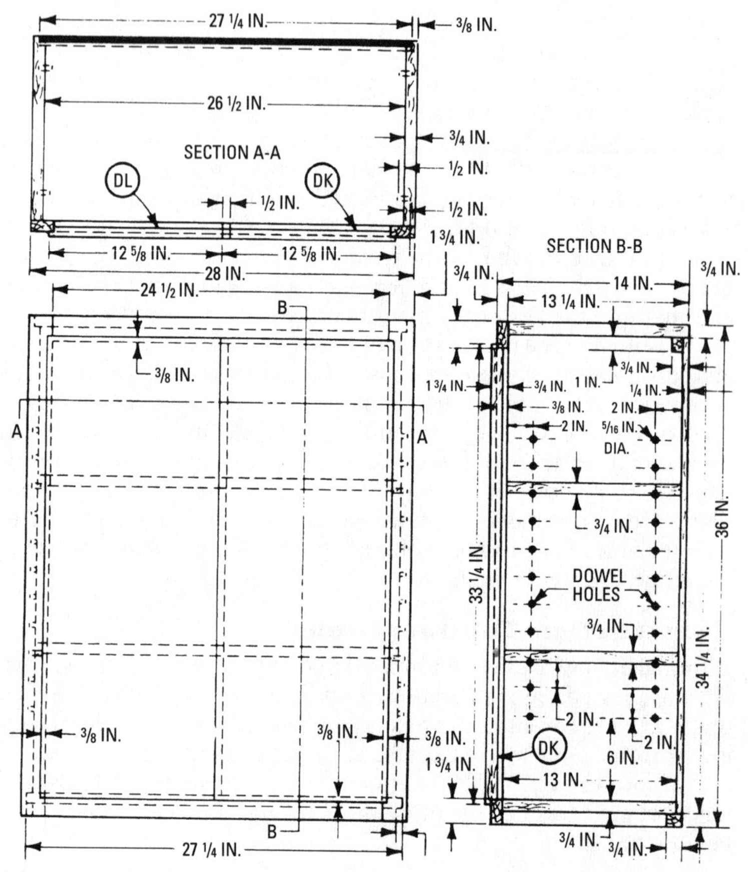

Figure 4-6 Construction details of a typical kitchen wall cabinet.

construction. The only difference is in the width and height. When the width and height of the cabinet have been decided, the material for each cabinet should be ordered. As the different pieces are cut, they should be marked with a key letter to facilitate assembly. If the cabinet is to be constructed of board stock rather than plywood or other wide materials, it may be necessary to glue up two or more narrow widths of stock to produce panels having the required widths.

The actual working drawings shown in the accompanying illustrations are based on a cabinet that is 28 inches wide, although the construction procedure will be similar for cabinets of any width.

Construction Details

When making a cabinet such as the one shown in Figure 4-6, it should be noted that actual construction is started with the side members. These pieces have a rabbet cut on the upper end to take the top piece, a rabbet cut along the back edge to take the back panel, a dado cut near the lower end for the bottom member, and two grooves cut at the upper and lower back edges to take the back rails.

After the necessary work on the two side members has been completed, the carcass for the cabinet can be assembled. When the carcass is completely assembled, the corners should be checked with a try square to make certain that they are square. A temporary diagonal brace may be fastened to the front edge to keep the unit square while the back panel is being applied.

The front frame is constructed of the two stiles and the bottom and top rails. The rails are joined to the stiles by means of mortise-and-tenon joints. Glue is spread over the joints, and the frame is set in clamps. The corners of the assembled frame must be checked to make certain that they are square. The shelves are cut to size in the regular manner, but they require the cutting of small circular grooves in each end to properly engage the supporting dowels. The door (or doors) should not cause any difficulty when assembling it to the unit if it is properly cut to fit the cabinet.

Varied and abundant types of cabinet hardware are available, as a glance through a hardware catalog shows. Various classes of hinges, door pulls, catches, and latches are generally found at any well-equipped hardware store. If catches are to be placed on double doors, then the strikes should be fastened to the underside of the shelf, to the top rail of the cabinet above the door, to the lower rail below the door, or to the bottom shelf. The strike and catch are usually provided with elongated holes to permit their adjustment after they have been installed. If a series of cabinets is being planned, the necessary hardware required may conveniently be purchased at one time, thus ensuring the perfect matching of each unit.

Cabinet Installation

Because of the considerable weight of a fully loaded cabinet (such as the combination unit shown in Figure 4-7), a great degree of care must be taken to ensure its proper installation. In all homes of frame construction, as well as in many other types, the inside walls

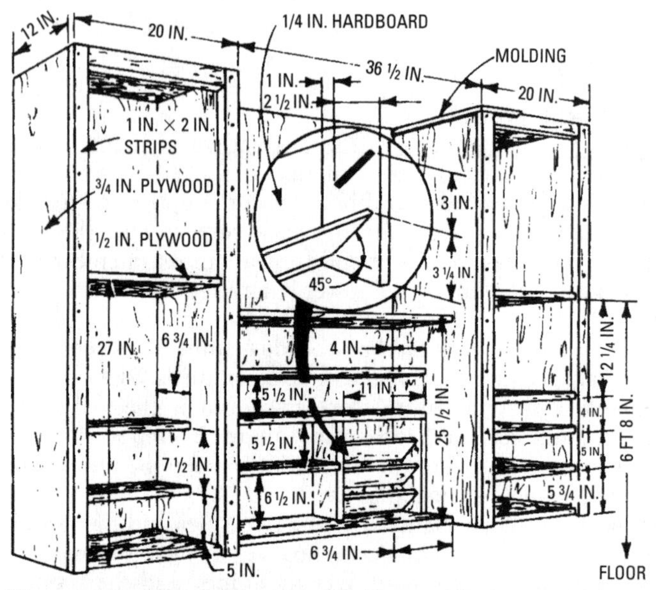

Figure 4-7 The layout of a wall-type kitchen-cabinet unit with center canned-food racks. Metal angles on the wall are used to support the cabinet.

are constructed on studs, which are on 16- (usually) or 24-inch centers. These studs will provide, in most cases, the only supporting means for the cabinet. The studs may be located by various means (such as by a fine drill or a thin nail. Be careful not to damage the wall during the stud-locating process. A plumb line may be dropped from the ceiling at the center of the studs to obtain the centerline of the studs throughout their entire length. Once the location has been determined, it is a simple matter to locate the holes in the top and bottom rails of the cabinet for final installation by means of suitable wood screws.

Use of a False Cabinet Wall

Most kitchen wall cabinets are designed so that the hanging wall units and full-length cabinets have a maximum height of 7 feet above the floor. The reason for observing this standard is for easier use, since any shelf placed higher than this distance cannot normally be reached without the aid of a stepladder. With cabinets installed at standard heights, however, there will usually be an open space near the ceiling where dust will tend to accumulate. A convenient method of eliminating this dust-collecting space consists of constructing a false wall extending from the top of the cabinet to the ceiling. When used, this method will add to the decor of the room and is adaptable to one or several cabinets as required.

To enclose the space above the cabinets, a framework must be installed (Figure 4-8). Sheetrock or plywood can then be attached to this framework. The stock used for the framework need not be of heavier cross-section than $1^1/_4$ inches × $2^1/_2$ inches or even lighter, depending on the bracing and height of the upright members. Cleats are fastened to the cabinet top and the ceiling to provide nailing strips for the uprights (Figure 4-8). Since the cleats are notched to receive the uprights, the panels may be fastened flush with the framework on all sides. Before nailing the cleats in place, however, the thickness of the panel must be determined, since the setback of the cleats that are attached to the cabinet top is controlled by this measurement. After having properly located the upper cleat, the next step is to locate the ceiling beams. If the ceiling beams run at right angles to the cleat, no difficulty will be encountered when fastening

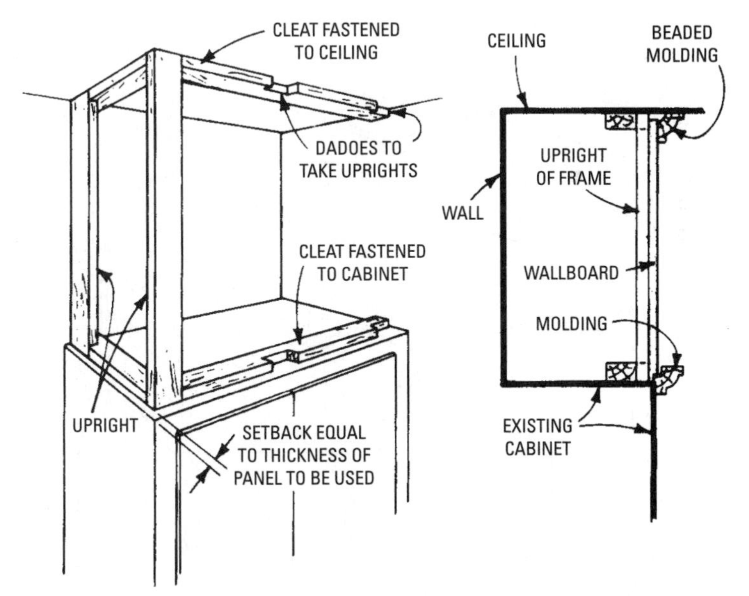

Figure 4-8 A simple construction method for enclosing the wall space above an existing wall cabinet.

the cleat to the ceiling. If, on the other hand, the beams are found to run parallel to the cleat, the nailing strips will have to be applied at right angles to the beams and extended out far enough to provide a nailing surface for the cleats.

After the framework has been nailed in place, all that is necessary is to cut the panels to size. These are usually fastened by means of fourpenny nails. When applying molding to the false cabinet wall, care should be taken to match the existing molding.

The next step is the sanding and finishing operations. The sanding removes whatever scratches are left in the wood. The false cabinets are then finished to match the original cabinets. If the paneled surface is to be papered, this work must be done before the molding is fastened.

Base Cabinet Construction

Figure 4-9 shows a typical kitchen cabinet of the base type. The cabinet is 24 inches deep from front to back and is 36 inches high. The cabinet, in most cases, is constructed of 3/4-inch plywood, although particleboard and flake board are sometimes used. The rails and stiles are usually cut from 3/4-inch plywood, although solid stock

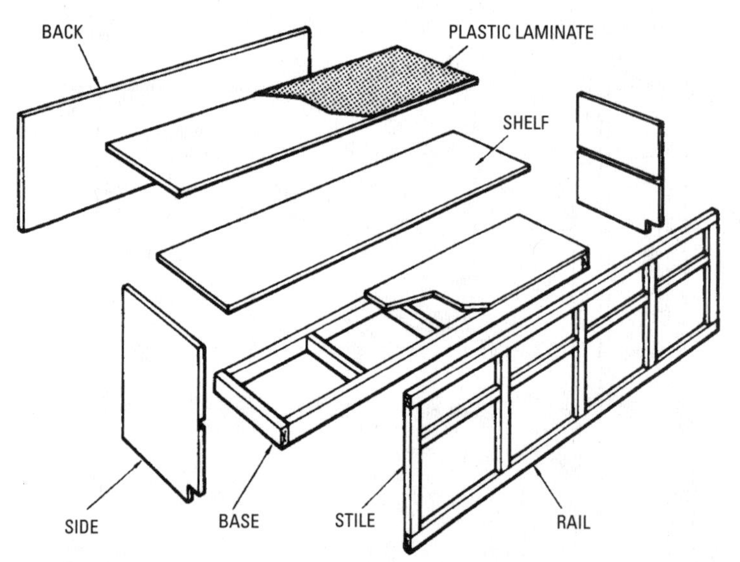

Figure 4-9 Construction details of a typical kitchen-unit base cabinet.

can be used. The sides of the base suit are dadoed to receive the shelf and plywood bottom. A $3^1/_2$-inch × $3^1/_2$-inch toe-space is provided at the bottom of the unit and 2 × 4 boards are placed under the plywood bottom. The sides of the base units are sometimes rabbeted to receive a $^1/_4$-inch hardboard back, or a 1-inch × 6-inch hanging strip is used in place of backing. All joints should be glued, using brads as required to anchor the work and prevent slipping.

Doors and Drawers

Two basic types of drawers and doors (flush and lipped) can be used. The *flush door* fits between the rail and stile and is very difficult to fit. The *lipped door* has a $^3/_8$-inch rabbet around the door, allowing for a margin of error. If a lipped door or drawer is used, the size of the front of the drawer or door is obtained by adding $^1/_2$ inch to the size of the opening.

Plastic Laminate Tops

The standard countertop these days is composed of a base of particleboard or plywood covered by plastic laminate, a high-pressure, extremely durable and hard material that resists moisture, scratches, and various other assaults.

Plastic laminate is available in a wide variety of colors, textures, and patterns (everything from plain colors to simulated wood). Textures may resemble many materials (including stone).

Installation of plastic laminate is relatively simple using contact cement, but unless you use the more expensive water-based cement, the job must be done outdoors. Regular contact cement gives off fumes that are highly combustible, even explosive.

Plastic laminate (which is $^1/_{16}$-inch thick and comes in large sheets) may be cut with a special cutter, hacksaw, fine-toothed saw, or coping saw. However, the easiest way to cut it is with a router equipped with a laminate cutting bit.

There are a number of ways to secure the laminate. A good procedure is as follows:

1. Cut the laminate about an inch wider all around than the base material (plywood or particleboard).
2. Use a brush or roller to apply contact cement to both surfaces. Make sure every square inch is covered.
3. When the laminate is just dry to the touch (a piece of paper will not stick to it), position it over the base material and lay a series of sticks across the material.

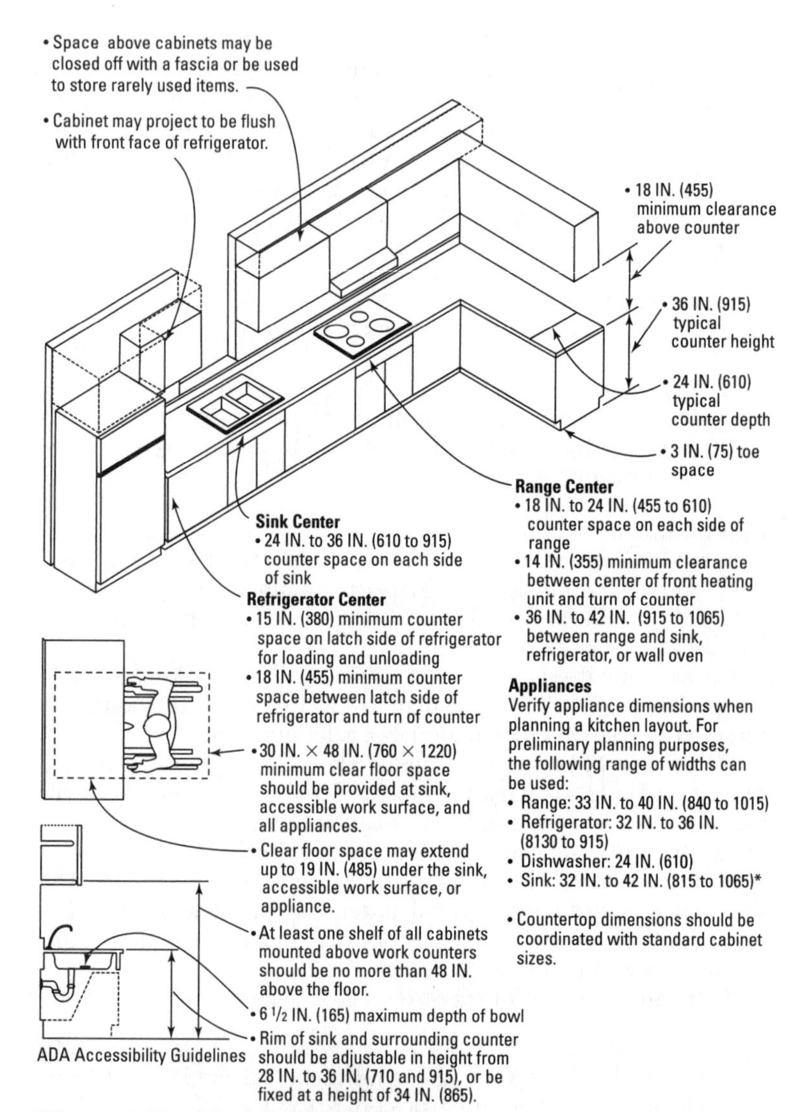

Figure 4-10 Kitchen dimensions

***Dimensions shown in parenthesis are metric and mm or millimeters.**

4. Lay the laminate in place and then, one by one, withdraw the sticks. As you do so, press the laminate in place, pounding it down with the heel of your hand so it sticks securely.
5. When all the sticks have been withdrawn, use a wood block to pound it down further.
6. Secure the edge strips, also trimming these to fit once they are secured.
7. Finally, use a wood block and a hammer to pound the laminate down. It's also a good idea to use a protective pad between the block and the laminate.

Kitchen Dimensions

Figure 4-10 shows a number of minimum dimensions the carpenter needs for a job in the kitchen. When working with laying out a kitchen, the spacing, height of cabinets, and placement of cooking devices (ovens, cooktops, and microwaves) are all important. Figure 4-10 illustrates how the architect views the kitchen in terms of the dimensions. These dimensions are accepted measurements. They will serve as a guide for those with questions on spacings and allowances.

Summary

The general number of cabinets required for the kitchen area depends on the size of the kitchen and the amount of cooking to be done. When remodeling a kitchen, there are certain limitations to observe (for example, the location of the sink and cooking range). The number and location of electrical outlets and fixtures are other important elements in kitchen planning.

Because of the considerable weight of a fully loaded cabinet, a great degree of care must be taken to ensure proper fit at all joints. Most kitchen wall cabinets are designed so that the hanging wall units and full-length cabinets have a maximum height of 7 feet above the floor.

In most cases, plastic laminate material is used for countertops. This material is generally 1/16-inch thick, resulting in a smooth surface. New countertops should be constructed of 3/4-inch exterior-grade plywood or particleboard, which should be firmly fastened to the cabinet.

To complete any cabinet, hardware of some type is required. Various classes of hinges, door pulls, catches, and latches are generally found at any well-equipped hardware store.

Review Questions

1. What determines the number of cabinets and the general placement?
2. What type of material is installed on countertops?
3. What grade of lumber is generally used in cabinet making?
4. What is generally done to the wall space above the wall-mounted cabinets?
5. What activities go on in a kitchen?
6. What is a work triangle?
7. A U-shaped kitchen surrounds the cook on ____ sides with continuous countertop and keeps walk-through traffic out of the work area.
8. Peninsulas can be used as ______ between the kitchen and adjacent room areas.
9. Most kitchen wall cabinets are designed so that the hanging wall units and full-length cabinets have a maximum height of ___ feet above the floor.
10. The standard countertop these days is composed of a base of particleboard or plywood covered by _______ laminate.

Chapter 5

Painting

Painting can be a deceptive trade. It seems like the techniques required are simple to master and, therefore, a typical painting job will be simple. However, such is not really the case. While the techniques are not complicated, using the right ones, along with using the right materials, can make the difference between having a job look very good or not good at all.

Tools and Techniques

A paint job requires special tools for application. The tools should be on hand and in good condition and ready for use.

Brushes

Classic large paintbrushes were called "pound brushes" by the professional, and were termed 4/0, 6/0, and 8/0 in size. They were made both round and flat, and the hog-hair bristles were bound with string or copper wire. The next smaller brushes were available in a dozen different sizes, and were called *tools* or *sash tools*. Some of these were bound with string, and others had their bristles fixed in metal. The smallest hog-hair brushes were called *fitches* and were used where tools would not fit. The smallest brushes of all were the camelhair brushes, made with long or short hair, according to the work to be done.

Modern brushes available to the consumer are made with flagged nylon or polyester bristles for use with latex paints and stains, or with natural bristle for use with oil-base paints and stains, varnish, enamel, or synthetic finishes. Nylon can also be used with oil-base paint. They range in size from 1/2 inch to 6 inches wide, have their bristles set in adhesive or rubber, and may have wooden or plastic handles. Their quality, aside from bristle composition, is determined largely by the length of their bristles (the longer the better), the number of bristles per square inch, endways (the denser the better), and the springiness or life of the bristles (not limp, but not so stiff as to leave tracks in the paint).

The use of a brush ensures good contact with surface pores, cracks, and crevices. Brushing is particularly recommended for applying prime coats and exterior paints.

In selecting a brush, you should choose one that is wide enough to cover the area in a reasonable amount of time. If you are painting large areas (such as exterior or interior walls, or a floor), you will want a wide brush, probably 4 or 5 inches in width (Figure 5-1). If

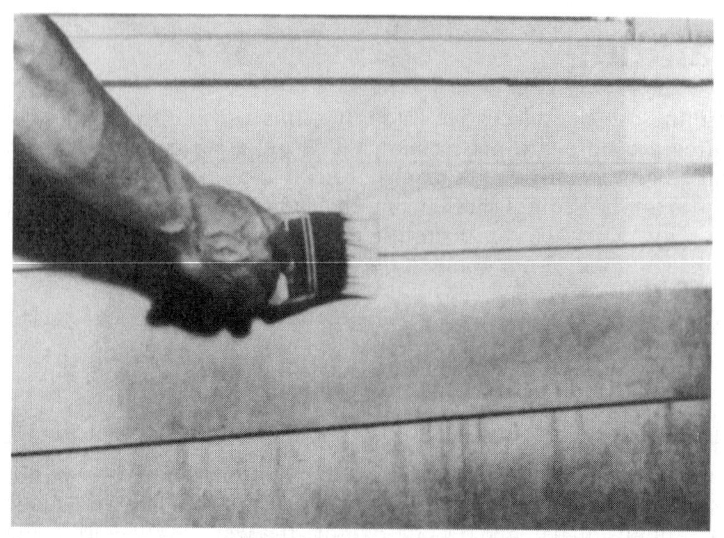

Figure 5-1 A 4-inch brush is right for wide, flat surfaces. *(Courtesy of Benjamin Moore)*

you are painting windows or trim, you will want a narrower brush, probably 1 to $1^1/_2$ inches in width.

The bristles should be reasonably long and thick so that they will hold a good load of paint. They should be flexible so that you can stroke evenly and smoothly. A medium-priced brush is the best investment if you do only occasional jobs.

Paint should be brushed up and down, then across for even distribution. On a rough surface, however, it is wise to vary the direction of the strokes so that the paint will penetrate thoroughly. Make your finish strokes straight up and down.

The brush should be held at a slight angle, and pressure should be moderate and even. Excessive pressure or "stuffing" the brush into corners and cracks can damage the bristles.

Always start painting at the top of a surface and move downward. For interior painting, do ceilings and walls first, then the doors, windows, and trim areas. If floors are to be painted, they should be done last. In addition, work toward the "wet edge" of the previously painted area (about 18 inches away), making sure not to try to cover too large a surface with each brush load.

Brushes in use should be dipped no more than two-thirds of their bristle length into the paint (Figure 5-2) and should be tapped on the inside of the paint bucket to remove excess paint instead of being

Figure 5-2 Brush should be loaded no more than two-thirds when painting. *(Courtesy of Benjamin Moore)*

stroked across the rim. Brushes in use for several hours should be washed out periodically during the day. This is to get rid of paint buildup below the ferrule. It also restores bristle springiness. If brush use is interrupted for short periods or overnight, the brush should be wiped off and the bristle end carefully wrapped in airtight plastic film to preserve the lay of the bristles and to prevent the paint on the bristles from drying until the brush is put back into use.

After the day's painting is finished, synthetic brushes used with latex should be gently (but thoroughly) washed in cool or lukewarm soapy water. Use a paintbrush comb to ensure that all bristles are separated and all paint is removed. Rinse the brush thoroughly to remove all soap, shake the water out of it, and shape the damp bristles into a wedge shape. Hang the brush up to dry. Brushes used in oil-base paint are treated the same way, but are thoroughly cleaned in mineral spirits to remove the paint before washing and drying. If the brush is not to be used for a long time, place it in its original wrapping (if available) and store in a cool, dry place (preferably hung up). Unwrapped brushes should be hung in a dust-free cabinet.

Brushes allowed to become stiff with hardened paint can be cleaned, but never seem to have the shape or feel they had before they hardened up. Oil-base paint can be softened by soaking the brush in linseed oil for 24 hours (or until it begins to soften), and then transferring it to a can of mineral spirits and working the paint loose the rest of the way by hand. If this conservative method doesn't work, try suspending the brush in a closed can, on the bottom of which is a shallow container of lacquer thinner (Figure 5-3). Fumes from the thinner will usually liquefy the hardened paint in the bristles, but will also strip the paint off the handle of the brush. Use a good liquid brush cleaner according to the directions on the can. This is a strong liquid with powerful fumes and potent action. Use it with care, and in a well-ventilated place. It will remove hardened latex as well as oil paint.

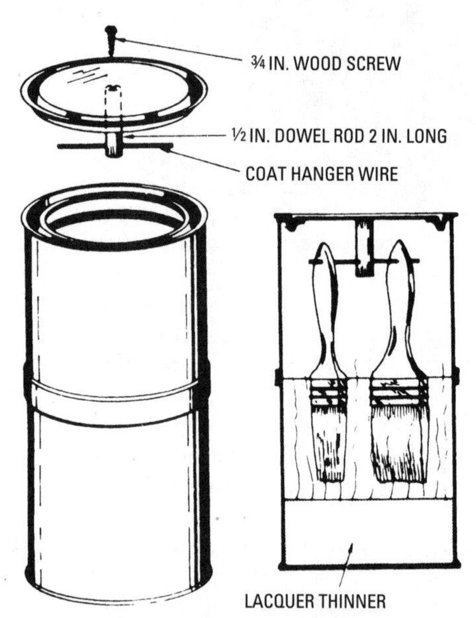

Figure 5-3 Suspending paint brushes above lacquer thinner in a closed can.

Rollers

The paint roller is an efficient, rapid means of applying paint. Almost anyone can get professional-quality results with practice and perfectly acceptable results with care, using either latex or oil-base paint. Rollers can be used to coat almost any material inside or outside the house.

The most popular consumer roller size is 9 inches, for walls, ceilings, and similar large surfaces, although a longer roller is available in some paint stores. The 7-inch roller is more economical with paint and easier to handle on beveled siding, fences, and smaller or cramped, flat surfaces. Besides paint, rollers can also apply glue, insecticide, contact cement, tile mastic, roof coatings, and blacktop sealer, among other things.

The first *roller covers* were made from lamb's wool, which is still used by some professional painters to apply oil-base paint. However, most are now made with man-made fiber with naps of assorted lengths. In general, the rougher the surface to be painted, the longer the nap of the roller should be to hold enough paint and to be able to reach the bottom of voids in rough surfaces. The roller packages will be marked as to the surface on which it should be used.

Roller frames (or the handle assembly on which the roller sleeve mounts) should be constructed of heavy wire for maximum strength at minimum weight, and should have a plastic handle molded in place (wood will loosen up and is too heavy). The cage over which the roller sleeve fits can be made of wire or flat steel, and should allow the sleeve to be slipped on and off easily. The end of the plastic handle should form a threaded socket into which an extension handle can be turned for ceiling or floor work.

A *roller tray* is necessary. It should be sturdily made and have a deep paint well. Liners are available so the same tray can be used for different colors of paints without cleaning it.

Pour paint into the tray until approximately two-thirds of the corrugated bottom is covered. Dip the roller into the paint in the shallow section of the tray, and roll it back and forth until it is well covered. If the roller drips when you lift it from the tray, it is overloaded. Squeeze out some of the paint by pressing the roller against the grid on the upper part of the tray above the paint line.

One roller-load of paint will cover an area about three roller-widths wide and four feet high. Roll it on easily in the pattern of a large "W," crisscross the strokes to cover the unpainted segments, and repeat on the adjacent wall areas until an area about 4 feet wide and the full height of the wall is covered. Then, go over this area lightly, using no more pressure than the weight of the roller, in one direction only. Resume painting the next area before the "wet edge" of the preceding area has dried, in order to prevent streaks and lap marks. Do ceilings and upper areas first, and then finish-stroke ceilings from the source of light rearward or parallel with the short dimension of the room. Lift the roller gradually from the wet surface as it nears the end of the stroke. A roller that is stopped and then lifted from the surface can leave a mark.

Rollers used with alkyd or oil-base paints should be cleaned with turpentine or mineral spirits. When latex paint has been used, soap and water will do a satisfactory cleaning job. If paint has been allowed to dry on the roller, a paint remover or brush-cleaning solvent will be needed.

Other Hand Tools

Sash brushes, *edging pads*, *trim rollers shields*, and similar devices are available to work the paint around trim, windows, corners, and ceiling lines when using a paint roller. They can be used before or after the main area is painted (sometimes both), but are more successful if used in advance. Figs. 5-4 to 5-8 provide some painting tips for brush and roller.

Sprayers

With the construction of modern equipment, spraying has become an increasingly popular method of applying paint to all types of interior and exterior surfaces. It is much faster than brushing, both in application and in the use of faster drying finishes. Small spraying units of the better types are very efficient, performing perfect work with paint, lacquer, varnish, and synthetics.

Airless Sprayers

For large-surface work both inside and out, the traditional compressed-air spray gun has been almost entirely replaced by the high-pressure *airless sprayer* that is capable of applying enormous quantities of paint with virtually no spray drift. So efficient is this machine that one worker can use it to paint the entire exterior of an unpainted brick ranch-style home with two coats of paint in one weekend, including cleanup time.

An airless sprayer consists of a motor that drives a pump that, in turn, sucks paint out of a reservoir and forces it through the nozzle on a spray gun. The motor may be electric, hydraulic, pneumatic, or internal combustion. The pump is a special reciprocating paint pump. The spray gun is equipped with a nozzle and valves made from tungsten carbide or similar material to withstand paint friction at pressures of 3000 psi or higher.

Nothing comes out of the spray gun but paint, which is driven to the surface to be painted in a precise, fan-shaped spray pattern. So precisely formed is this pattern that at a distances of less than

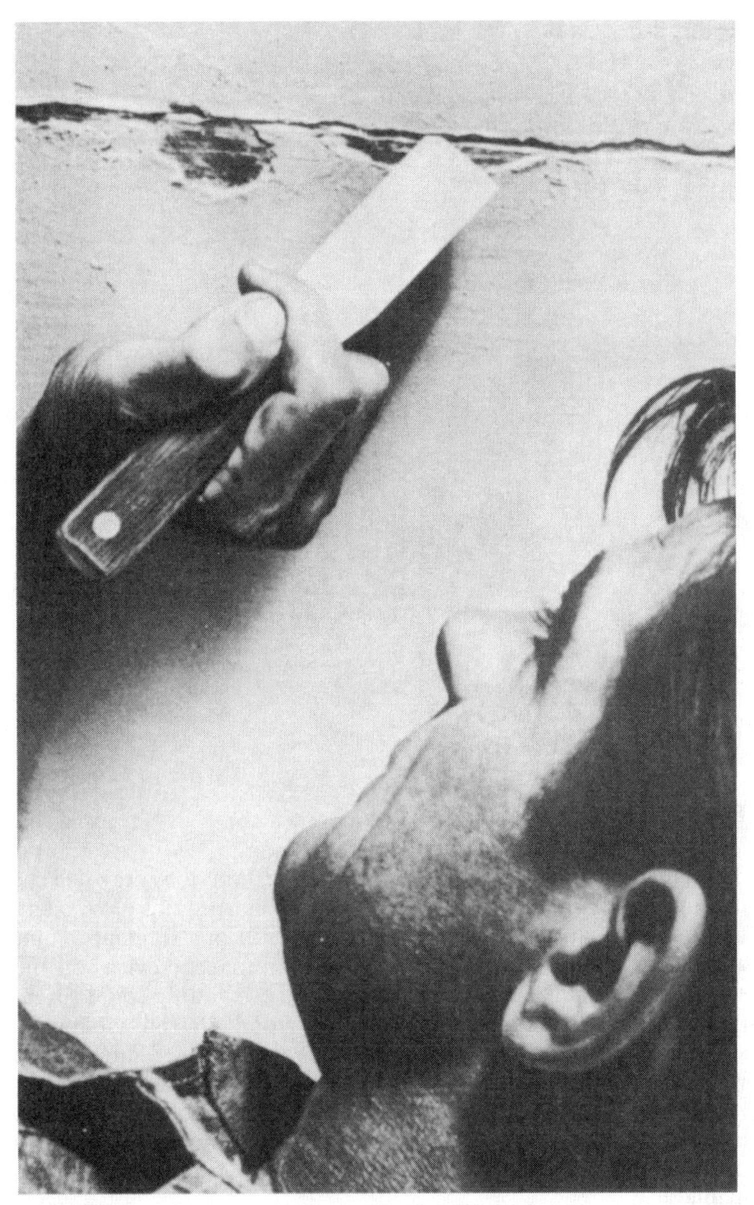

Figure 5-4 Patch and smooth before painting.

Figure 5-5 Cut in ceiling... *(Courtesy of Benjamin Moore)*

3 or 4 inches from the nozzle of a commercial unit, the spray will cut through bare skin like a knife. For this reason, the guns have safety tip guards or similar devices to help prevent serious accidents. Paint applied by airless spray will penetrate cracks and crevices on the surface to be painted that no other method of painting could reach, making it ideal for rough masonry or wood as well as smooth surfaces. The paint will not bounce off the surface. Paint is not wasted through drifting, as is the case with compressed-air sprayers.

Most airless sprayers are rental items because their high price makes purchase practical only for painting contractors or building maintenance people. Small, relatively inexpensive units are now available.

Airless units do not have a paint cup on the spray gun, as in compressed-air sprayers. Paint under pressure is transported

Figure 5-6 ...then roll.

through a single hose up to 50 feet long (or longer) from the pump to the spray gun. The pump unit, which is easily portable, can mount directly on any 5-gallon paint bucket, sit on the ground, or be designed with wheels (similar to a hand truck). In the latter instances, the pump draws paint from a separate container through a special suction hose or rigid pipe. Cleanup is simple and economical, since paint in the hose and most of the cleaning solvent (if used) can be recovered for later use. Almost any kind of paint can be applied to almost any surface by airless spray, but the use of lacquer is not recommended.

Compressed-Air Sprayers

These are traditional spraying units, which consist of an air compressor, spray gun, and connecting hose.

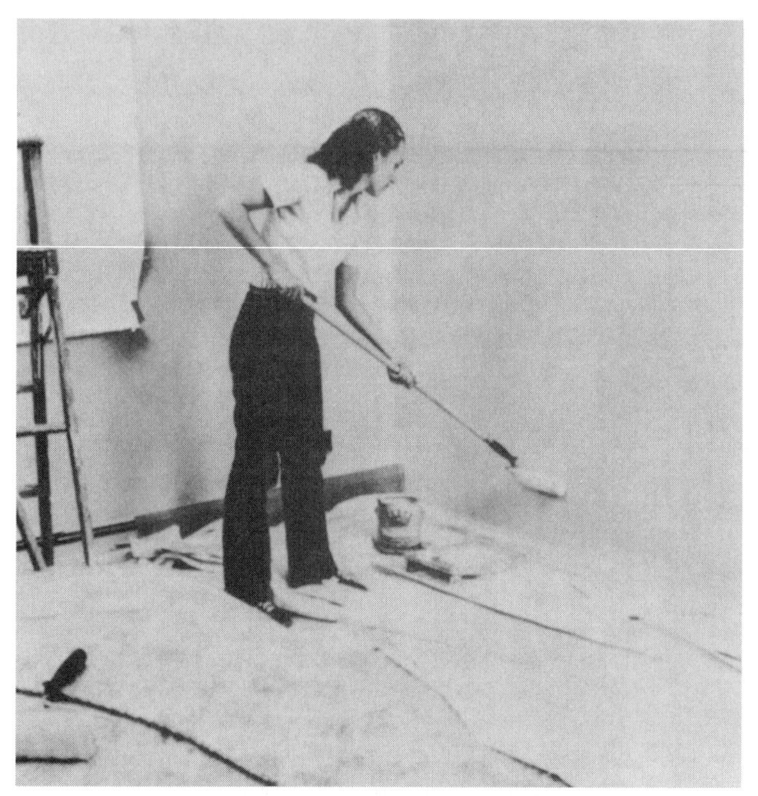

Figure 5-7 A long stick with a roller can make painting simpler. Do walls after ceiling. *(Courtesy of Benjamin Moore)*

Air Compressor

There are two kinds of *air compressors*: the piston type and the diaphragm type. The *piston type* consists essentially of a metal piston working inside a cylinder, very much like the piston in an automobile engine. On the down-stroke of the piston, air is drawn into the cylinder and on the up-stroke, this air is compressed. Figure 5-9 shows a piston-type stationary paint sprayer, and Figure 5-10 shows a portable type.

The *diaphragm-type compressor* differs from the piston type mainly in that it is equipped with a rubber diaphragm taking the place of the piston. Because of the fact that the diaphragm type of compressor is rather inexpensive and requires less maintenance, it is usually preferred in the spraying applications that are performed around the home where the air requirements are small.

Figure 5-8 Painting window sash with brush at the angle shown makes the job easier.

The motor supplying the power for the compressor is furnished in fractional horsepower, and it determines the size of the compressor. For home spraying, $^{1}/_{4}$ to $^{1}/_{3}$ horsepower is the smallest motor practical to use for satisfactory spraying. Most compressors should be able to deliver a constant-volume supply of air to the gun at a pressure of about 40 pounds per square inch (psi).

Spray Guns

In operation, the *spray gun* atomizes the material that is being sprayed and the operator controls the flow of the material with the trigger attachment on the gun (Figure 5-11). Spray guns are supplied with various sizes of nozzles and air gaps to permit the handling of different types of material, as shown in Figure 5-12. An air adjustment built into the gun makes possible a change in the shape

Figure 5-9 Typical 1/3 -horsepower stationary paint sprayer.

Figure 5-10 A portable 1/2-horsepower paint sprayer.

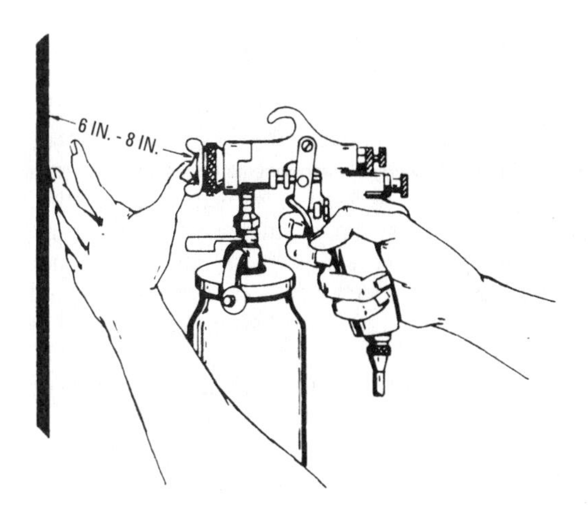

Figure 5-11 A typical paint spray gun.

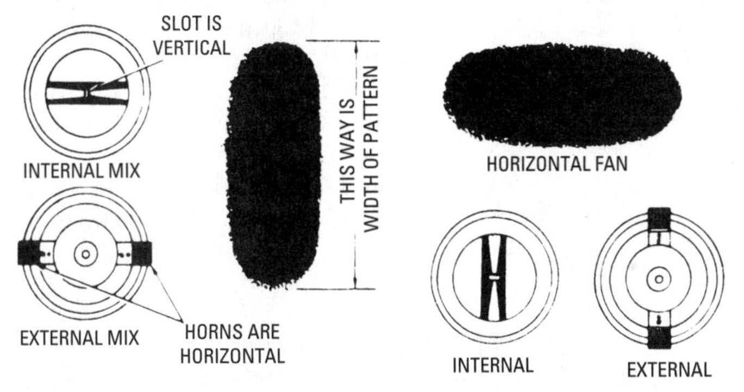

Figure 5-12 Various nozzles used for different patterns of spray.

and size of the spray pattern, ranging from a wide fan-shaped spray to a small round spray, as the occasion requires.

Spray guns are either *external mix* or *internal mix*. In the external-mix type of gun, the material leaves the gun through a hole in the center of the cap, while air from an annular ring surrounding the fluid stream (as well as from two or more holes or orifices set directly opposite each other) engages the material at an angle. Spray guns in which the air and material are mixed before leaving the orifice of the gun are of the internal-mix type. The suction-feed gun, as the name implies, siphons the material from a cup attached to the gun and is generally used for radiators, grills, and other small projects that require light bodies and only a small amount of material.

The pressure-feed gun forces the material by air pressure from the container. Air directed into the container puts the material under pressure and forces it to the nozzle. If compressed-air spraying is employed, the pressure-type gun is undoubtedly the best all-around gun for use with small compressors and with such materials as house paint, floor enamels, wall paints, and the like. Any spray gun is either a bleeder or a nonbleeder type. A *bleeder type* is one that is designed for the passage of air at all times, whereas a *nonbleeder gun* cannot pass air until the trigger is pulled. This nonbleeder type of gun is used only when air is supplied from a tank or from a compressor equipped with pressure control. Bleeder guns are always used when working directly from small compressors (such as those used in small workshops, or for spraying jobs around the home).

Air Compressor Accessories

One of the most useful accessories for home-use spraying is the *angle head*. An angle head fitted to the pressure gun prevents excessive tilting of the fluid cup when spraying floors, ceilings, and similar surfaces. A respirator is seldom required for work of short duration, but it is a very useful item when spraying for long periods in confined quarters.

Another useful accessory is the *condenser*, which filters water, oils, and various foreign particles out of the air supply. If the condenser is fitted with an air regulator, the combination is termed a *transformer*. With a condenser-regulator combination, it is possible to have the air filtered, while the regulator allows setting the air pressure at a predetermined rate as required for the particular job. Additional useful accessories consist of an *air-duster gun* and extra lengths of *air hose* with couplings to facilitate an increase in the working radius.

Compressed-Air Spraying Technique

Prior to commencing the actual spraying job, it is good practice to test spray a few panels to obtain the feel of the gun and to make any necessary adjustments. Such practice spraying can be done on old cartons or on sheets of newspaper tacked to a carton or box.

With the spray cup filled with water-mixed paint or other inexpensive material, practice spraying can be done directly on any scrap material, as previously noted. It will be useful to experiment with the full range of fluid adjustment, starting with the fluid needle screw backed off from the closed position just far enough to obtain a small pattern an inch or so wide when the trigger is pulled all the way back. The spray-gun stroke is made by moving the gun parallel to the work and at right angles to the surface. The speed of stroking should be about the same as brushing. The distance from gun to work should be between 6 and 8 inches, as shown in Figure 5-13.

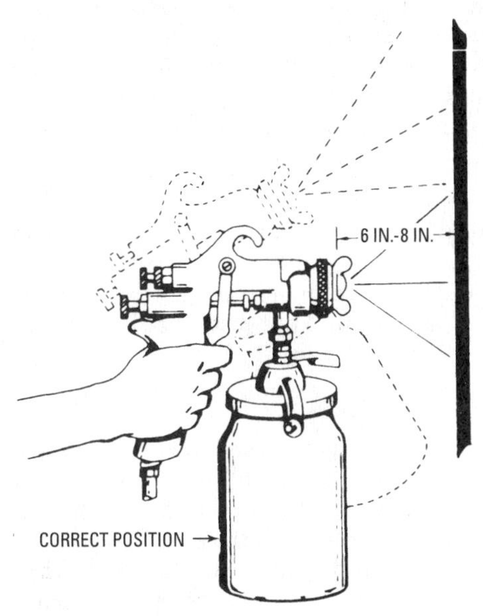

Figure 5-13 The correct position of spray gun when spraying.

Practice should be done with straight uniform strokes moving back and forth across the surface in such a way that the pattern laps about 50 percent on each pass. An appreciable variation of distance between gun and work (or angle of the gun) will result in uneven coatings. Therefore, always hold the gun without tilting and move it as described. Spraying is merely the action of the wrist and forearm and the proper technique is easily acquired. Fairly large pieces of scrap wood remaining from the project are excellent for practicing spray patterns, because this gives the operator an idea of any problems to expect in actually spraying the final project itself.

Release the trigger at the end of each stroke to avoid piling up material. Avoid pivoting or circular movements of the wrist or forearm or any oblique spraying that will cause material to rebound from the surface and produce excessive mists and waste material as well as a patchy, unsatisfactory job.

Cleaning the Gun

To obtain continued and satisfactory service, a spray gun (like any other tool) requires constant care. It should not be left in a thinner overnight, because such practice removes the lubricant from the packing through which the needle moves, thereby causing the needle to stick and the gun to "spit." A dirty air gap causes a defective spray. Thus, it is good practice to clean it promptly after the spraying job is finished. At the end of the day, remove the gun from the cup and hose, and thoroughly clean it. Clean the entire gun, the outside as well as the inside. The cup or tank should likewise be cleaned.

Materials

Knowledge of the materials of which paints are made can help you choose the correct product, use it in the right way, and solve some of the problems that arise in their use. To understand these materials, we will look at them very closely to find out just what happens as they are used.

The Coating System

Paint and other finish products (such as varnishes and stains) must meet a wide range of requirements. They should not be too costly to make and must remain in a liquid, usable condition until bought and used. They should spread easily onto a surface and flow into a flat layer, and stop flowing just when they are smooth. Finally, they must harden into a tough durable film that is easy to renew when it wears out.

Liquid paint is made of three very different materials. The *pigment* gives color to the paint and is mixed in the *vehicle* (or liquid part) of the paint. The vehicle is made of two parts: The *binder* cements the pigment particles together and sticks it to the surface being painted, and the *solvent* thins the binder so the paint can be easily spread. Sometimes the binder is thin enough that solvent is not needed. Clear finishes (such as lacquer and varnishes) are simply binder and solvent without pigment.

All three parts work together as a system to make the final film possible (Figure 5-14). Solvent, binder, and pigment are mixed to make the liquid as you see it in the can. When you spread the paint, its surface area exposed to the air is increased. The solvent begins to evaporate, and the film feels tacky and then dry to the touch. Even though the film feels dry, it may not reach its full strength until after some days or weeks of curing.

After the film has cured, it begins its life of protecting the surface. You may have noticed that a recently painted surface will begin to

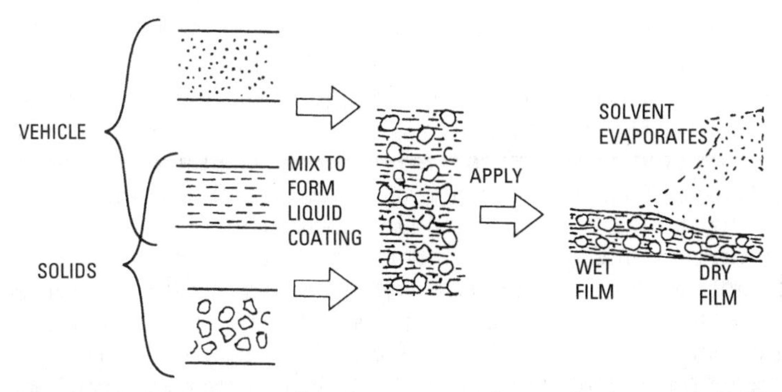

Figure 5-14 Generalized composition of pigmented liquid coatings and manner of film formation for solvent-based coating.

collect dirt and look soiled. After a while, the glossy sheen will be weathered away along with the dirt and the paint will appear clean again. The film gradually wears away and will have to be recoated before it is completely gone. If the film fails before this happens (perhaps by cracking or peeling), renewal will be much more difficult.

There are two main types of film forming coatings: *emulsion-based paints* (latex) and *solvent-based paints* (enamels, lacquers, and varnishes). Solvent-based paints are sometimes referred to as *oil paints*, but there are other types of binders as well.

Solvent-Based Paint

The binders of oil finishes and pigmented oil paints are made of liquid vegetable oils (such as linseed and tung oil). After the initial cure, an oil paint film is flexible, yet tough and durable. However, the film continues to harden, until in a few years it becomes brittle and begins to fail.

This continued curing is the cause of "chalking paint." As the binder fails at the surface of the film, its grip on the particles of pigment is released. Chalking can cause white streaks below as it is washed from the surface of a wall. Chalking paints should not be used above masonry or paint of contrasting color, where streaking would be a problem. It is, however, an ideal way for a paint to fail, because as the top surface of the film is slowly eroded away, removing dirt and grime along with it, the base of the film remains firmly attached to the surface. It can then be easily recoated.

The *alkyd resins* appear in most of the common types of solvent-based coatings. Alkyd resin paints dry in less time than oil paints.

The short drying time allows only limited penetration, but the resin is highly adhesive and the film bonds well with the surface being coated. This makes alkyd coatings very good for new work, where adhesion and film strength are important, but it can cause problems in recoating old paint films that may be weak. Continued curing of the stronger alkyd film can cause shrinkage that can eventually pull the lower, weaker layer apart to cause cracking and peeling.

Primers and undercoats made with alkyd binders seal the surface and "hold out" or do not penetrate. They are useful in covering surfaces that vary in porosity (such as wallboard with compound on the joints). On this kind of surface, oil paints would soak in at different rates and there would be unsightly differences in the surface sheen.

Interior white or light-colored topcoats made with alkyds are less likely to yellow with age than oil paints. They are most resistant to this discoloring when made with safflower, soya, or refined tall oil.

Alkyd resins react unfavorably with alkali. Coatings containing large proportions of them should not be used directly on plaster or masonry.

Emulsion-Based Paint

Commonly called *latex paints* because early types used resins made from latex rubber, the emulsion-based paints work on a very different principle from the solvent paints such as oil and alkyd. *Emulsion* is defined as a suspension of fine droplets of liquid dispersed in another liquid. In the emulsion paint system, the vehicle is made of a thick viscous resin in the form of microscopic spheres that are suspended in water. This emulsion is a brushable mixture that has the pigments dispersed in it. The pigments are the same as those used in solvent paints. The emulsion system allows the use of resins that are much thicker in consistency than those used in solvent coatings, while letting the paint as a whole remain spreadable. The thicker resins are more resistant to the chemical attacks that break down a paint film.

As soon as the paint is spread on a surface, the water begins to evaporate (Figure 5-15). After a few hours, the water is mostly gone and the individual spheres of resin begin to merge into the final film, trapping the particles of pigment.

This type of film is cohesive (or strong within), but is not very adhesive. It will not penetrate the surface, requiring closer attention to surface preparation and priming to make it bond. Oil and alkyd primers are often used with latex paints, but the manufacturer's recommendations should be followed as each product is designed

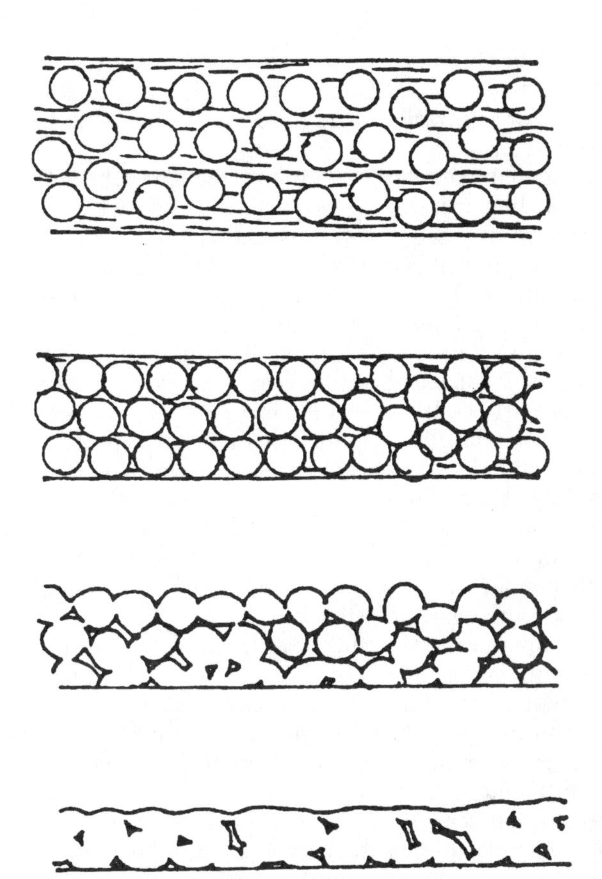

Figure 5-15 After evaporation of water, tiny spheres of latex resin coalesce to form the film (pigment left out for clarity).

to work best with a certain primer. Don't apply latex topcoat paint directly to an unprimed wood surface. The water in the vehicle can cause the grain to rise, and may stain the latex film if there are water-soluble stains in the wood. Redwood and cedar contain such stains. If glossy surfaces of old paints are not sanded and primed, the film will shrink or "crawl," leaving irregularly spaced gaps in the film.

When the latex film dries, it forms voids that allow it to be somewhat permeable to moisture. This can be an advantage on exterior wall surfaces where water vapor movement from the interior of the wall to the outside is desirable. Some of this advantage is lost with each additional coat.

If this paint is applied when the temperature is too low (below 50°F), the binder may not flow together and merge. The film will dry, but the particles of resin will brush off the wall along with the pigment as a fine dust.

Since water is essentially the entire vehicle, it is the only substance evaporating from the film as it dries. The user is not exposed to the potentially harmful vapors used in solvent-based coatings, making this one of the safest kinds of paints to use in areas with poor or no ventilation. Cleanup with water is certainly safer than with the solvents needed for other paints.

All three main types of resins used in emulsion paints produce a coating that is easily applied, has low odor, and is rapid drying. *Butadiene-styrene* (BDS, latex) was the first resin used in latex paints. It forms a less flexible film than the other resins, but is sometimes still used for interior products. *Polyvinyl acetate* (vinyl) has been more commonly used for interior work in recent years. *Acrylic* resin is the most resistant to alkali and water, which makes it good for masonry and interior areas that need to be washed frequently. The tough flexible film it forms is also useful on exterior wood that expands and shrinks with changes in the weather.

Exterior latex paints are by far the easiest, fastest, most convenient, and most popular paints to use. They are tough, long lasting, quick-drying, mildew-resistant, and allow easy tool and splatter cleanup with soap and water. When dry, the paint repels rain, hose water, dew, and other water, but allows water vapor from inside the building to pass through to the outside air without peeling.

Technological development in these paints has been rapid, and the most widely available now is the acrylic latex. This paint is available under a number of brand names and formulations, all of which have the characteristics mentioned in varying degrees, and a flat or slightly glossy, nonchalking finish. Some need no primer over new wood, but most do require one for best adhesion and prevention of bleed-through.

Two finish coats are usually ample, but they must be applied as full coats and not brushed or rolled out thinly (they go on with so little drag on the brush that the tendency is to brush them out too much). Allow an hour between coats or a little more if the humidity is high, even though the paint will dry in 20 to 30 minutes. Acrylic latex can be applied over damp surfaces with no problem, but not after or during a heavy rain. To avoid damage to the paint film, stop painting in time to let the paint dry before it is struck by rain or heavy dew. Bonding and curing of the paint is adversely affected by relatively low temperatures, and almost all manufacturers specify that the paint not be applied if atmospheric or surface temperatures are 50° or lower. Apply the paint as it comes from the can, even if it seems too heavy. Some brands are almost jelly-like, and all are thicker than oil-base paint. Thin it only for spray application, and then according to instructions on the can. Generally, spray thinning should amount to no more than 1/2 pint of water per gallon of paint.

Acrylic latex comes in a number of stock colors that vary from brand to brand, and most paint and hardware stores will custom-mix special colors for a small charge. Glossy trim paints (which are also acrylic latex) are available in matching or contrasting colors.

For best results (and to save time, work, and money), apply latex only over latex or oil only over oil unless previous coats are stripped to bare wood.

Many of the qualities mentioned for exterior latex also apply to its interior counterpart. It is quick-drying, easy to apply, shows no lap marks, is available in literally hundreds of shades and tones of color, is durable and washable, and has an odor not usually regarded as objectionable (but it should be applied and allowed to dry with plenty of ventilation). It is by far the most popular interior paint sold, and is available in flat and semi-gloss.

Like exterior latex, the interior variety should be used as it comes from the can, without thinning. Some brands are called "drip-less" latex and are heavy-bodied to prevent brush drips and roller splatter. One manufacturer's advertising shows gel-paint to be so thick that holes could be scooped into a can of it, yet it was just right for application.

New work requires at least two coats of paint. After that, depending on the color, you may be able to use only one. Keep in mind that yellows have the poorest hiding power. Blues have the best hiding power. If you decide you want white ceilings, use the latex ceiling paint. It has extra pigment for good hiding power, and can cover a previously painted ceiling in one coat. If it is unavailable, have your paint dealer add just a small amount of his darkest blue tinting color to his densest tinting base. The result will be indistinguishable from pure white, but will cover very well.

Low Volatile Organic Content (VOC) Coatings

Volatile organic contents (VOCs) are the solvents in paints and other coatings that combine with oxygen in the air to create low-level ozone, the main ingredient of smog. A few states have set VOC limits to protect the environment. Many more will set limits within the next few years.

Paint manufacturers have dealt with the new regulations in two ways. They are expanding and heavily promoting their lines of emulsion-based coatings. In addition, they are changing the formulations of their solvent-based coatings in three important ways:

- Removing solvents and adding binders and pigments makes "high-solids" coatings. These paints are so thick and heavy that it is difficult to spread them out thin enough. Application takes more time and effort.
- Using low-viscosity resins for binders that require less solvent make the cured paint film less durable and short lived.
- Using solvents that are more environmentally acceptable but also more expensive.

Application methods and performance of these paints are substantially different from the common solvent-based paints. Wet paint films may skin over and feel dry, but still be liquid enough underneath to slide right off the wall if you lean against it. Emulsion-based coatings usually take more time to achieve top-quality appearance and performance. Surface preparation is particularly important because of poor penetration and adhesion.

It is a period of great change in the paint industry. Formulations can change from month to month. Major manufacturers are pulling out of whole states at a time, or dropping whole lines of products, so you cannot depend on having tried-and-true products to use. So, what can you do to prepare? Begin experimenting with the new formulations now so it is not a sudden shock when you are forced to switch. Set up your own test panels and evaluate performance over extended periods, such as a year or two. Use the new paints on smaller jobs, so if there are problems at least they will be small ones. By the time changes are mandatory, you will have the experience and knowledge to make the shift.

The best way to track changes in the regulations for your area is to contact your state environmental protection agency for changes in legislation. In addition, keep in touch with state contractors' associations to see how strictly new laws are enforced.

Estimating

You can roughly figure the amount of paint you need for flat surfaces by simply multiplying the height (or length) times the width and dividing the result into the coverage estimate on the label. If, for example, you want to paint a room that has 416 square feet of wall area, a gallon of paint advertised as covering 500 square feet will be adequate for one coat. Second coats generally require less paint.

However, these are only average estimates of coverage. Some surfaces are more absorbent than others.

Colors

The average paint store has paint-mixing machines in which, at least theoretically, they can match any color you bring into the store. That is not true. Different surfaces absorb paints to varying degrees, so no matter how close the paint dealer can get, the paint will never be exactly the shade that you want.

At any rate, buying paint custom-mixed is expensive. It is better for the painter to know how to mix colors. You can get exactly the shade you need, and for only the price of the colorant.

Colorant for the job can be *universal tinting colors*. These are so-named because they are good for both latex and oil-base paints. They come in tubes for small jobs, and in cans if you expect to use the particular color frequently (Figure 5-16).

Figure 5-16 Tinting colors can create any color you wish. *(Courtesy of Benjamin Moore)*

When using colors, the following tips should be helpful:

- Add the color to the paint gradually, mixing it in thoroughly.
- The colorant will mix in more readily if it is first mixed with a tiny bit of thinner.
- Try the mixed paint on a spot where it is in full light. Shadows can create false readings.
- Let the paint dry before determining its exact shade.
- Do not add more colorant than specified on the container.

Figure 5-17 shows the three primary colors, their secondaries, and what may be called *tertiary colors*. Opposite each of these there has been placed one of the notes of the chromatic music scale forming a perfect octave. It is interesting to note that the claim has been made (and with much insistence) that any scheme of color that may be selected and that may be struck as a chord will, if the chord is harmonious, become a harmonious scheme of color. If this chord produces a discord of music, there will be a discord of color.

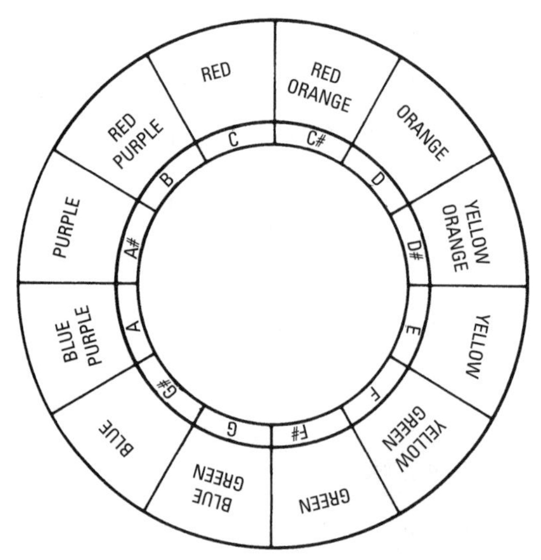

Figure 5-17 The three primary colors with the secondary and tertiary colors shown.

By definition, *primary colors* are those that cannot be made by mixing two or more colors together. The three primary colors are *red, blue*, and *yellow*. The colors obtained from mixing any two of the primary colors are called *secondary colors* (Figure 5-18). The three secondary colors are *purple, green,* and *orange*: Combining red and blue gives purple; blue and yellow gives green; and red and yellow gives orange.

By mixing any two of the secondary colors together you get what are called tertiary colors, which are *citrine, olive*, and *russet*: Combining orange and green gives citrine; green and purple gives olive; and orange and purple gives russet.

Black and white are not regarded as colors. A good black can be produced by mixing the three primary colors together in proper

- Ivory shades (white, lemon chrome, or ocher)
- Old rose (white, ocher, Venetian red, or pure Indian red and black)
- Nile blue and Nile green (white, Prussian blue, lemon chrome)
- Light citrine, light olive, light russet

For ceilings, the best tints are the creams and ivory tints, and gray. Creams and ivory tints are made from white tint with one or more of the following colors:

- Lemon chrome
- Orange chrome
- Ocher
- Raw sienna

To produce a warm tone, add a small quantity of burnt sienna, vermilion, or Venetian red. To produce a colder tone, use a little green, black, raw umber, or blue. Grays are made from white tint with black, black and green, blue and umber, black and red, red and blue, or burnt sienna and blue. Light colors are used for ceilings in preference to dark colors. Contrasting colors are better for ceilings than a lighter tint of the wall color.

Surfaces

When selecting paints, you must be aware of the individual characteristics of a variety of surfaces, including plywood, woods, and wood floors.

Plywood

The painting and finishing characteristics of plywood are essentially those of the kind of wood with which the plywood is faced. In large sheets of plywood, any wood checking is more objectionable in appearance than it is on boards of lumber. Plywood with hardwood faces usually is not much more prone to checking than lumber of the same species, but plywood faced with most softwoods is likely to check to some extent even when used indoors with protective finishes. Such checking is more conspicuous with some finishes of light color, but somewhat less as the gloss diminishes and as the color is darkened greatly.

With opaque finishes, the checking is least objectionable when flat finishes with rough surfaces are chosen (such as stippled, textured, or sanded finishes). Checking is less readily observable with natural finishes than with opaque finishes, because the grain pattern of

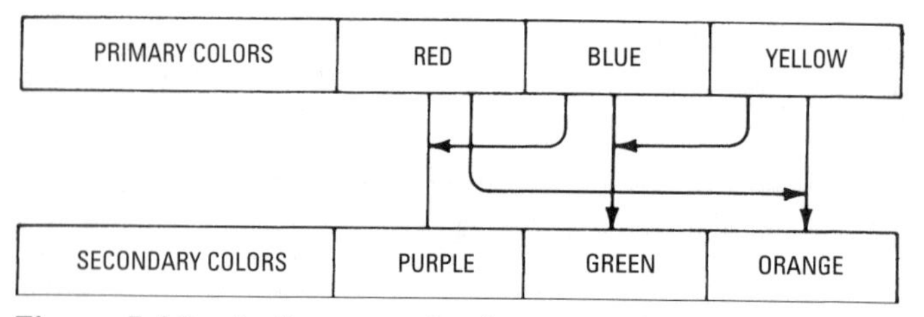

Figure 5-18 A diagram of primary and secondary colors.

proportions. By adding white to any color, you produce a *tint* (that color. By adding black to any color, you get a *shade* of th: color. That is the difference between tint and shade. The use c black subdues (or lowers) the tone of any color to which it is added To preserve the richness of colors when you want to darken them use the primary colors instead of black. To make a yellow darker use red or blue, and to darken blue, add red and yellow, and so on.

Every shade or tint of color required by the painter can be made from red, blue, and yellow with black and white. To make any of the umbers or siennas lighter in color, and to preserve the clear richness of tone, always use lemon chrome instead of white. Use the white if you want a subdued or muddy umber or sienna color.

The most useful primary colors are the following:

- *Yellows*—Lemon chrome, deep ocher
- *Reds*—Vermilion, Venetian red, crimson lake
- *Blues*—Prussian blue, ultramarine

Yellow ocher is a useful color. In its pure state, it is admirable for large wall spaces. If you are in doubt as to what color to use to complete a color scheme, you will find that ocher or one of its shades or tints will supply the missing link.

Red paint on walls makes them look smaller and will absorb light. Yellow gives light and airiness to any room and it will also reflect light. Useful colors in large quantities for churches, public halls, and so forth, are the following:

- Primrose red
- Terra cotta (white, burnt sienna, lemon chrome)
- All tints of ocher
- Flesh colors (white and burnt sienna)
- Pea green, apple green
- Gray green (white, Paris green and a touch of black)

the wood distracts attention from the checking. Softwood plywood may be used without danger that checking will mar the finish if the exposed face of the plywood is covered with paper, cloth, or resin-impregnated paper firmly glued in place. For exterior exposure, of course, the glue must be thoroughly weather-resistant.

Woods

Some raw woods allow a wide choice in paints and painting procedures, whereas others are more exacting in their requirements. Woods differ also in the appearance of natural finish obtainable with them in the ways in which they behave if exposed to the weather without protective coating or treatment. Woods that offer the greatest freedom of choice for painting are the woods that are low in density, slowly grown, cut to expose edge grain rather than flat grain on the principal surface, either free from pores or have pores smaller than those in birch, and free from defects (such as knots and pitch pockets). Flat-grain boards hold paint better on the backside than the pitch side. All the pines, spruces, and Douglas firs are improved for painting by kiln-drying to fix the resin in the wood so it will not exude through the coating.

Eastern red cedar and Spanish cedar contain volatile oils that prevent proper drying of most paints and varnishes, unless most of the oils are removed by exceptionally thorough kiln-drying. Incense cedar, port oxford cedar, and cypress may sometimes give similar troubles if they contain too much of their characteristic oils. The method of seasoning on most other woods has little or no effect on "paintability." Water-soluble extractives in redwood, Western red cedar, walnut, chestnut, and oak may retard drying of coatings applied while the wood is damp or wet, and may discolor paint if the wood becomes thoroughly wet after painting. Otherwise, such extractives contribute to better durability of paint.

On exterior surfaces, softwoods that expose wide bands of summerwood require extra care in painting. This is because it is over such bands of summerwood that the paint coating, when embrittled by the action of the sun and rain, begins to wear away by crumbling or flaking. Repainting is usually considered before too much summerwood becomes exposed. Coatings on interior surfaces seldom wear away by crumbling or flaking. However, greater smoothness of painted surfaces may be demanded on interior surfaces. The presence of wide bands of summerwood makes it necessary to take extra care in smoothing the surface before painting and in applying undercoats to keep the wood from showing ridges in the coating (commonly called *raised grain*).

Sometimes the easiest woods to paint smoothly may be considered too soft and too easily dented to be a good choice for interior woodwork. In these cases, a compromise must be reached between the properties that favor smooth painting and those that give enough hardness for the intended use. For good service of paints or other finishes, wood must be kept reasonably dry, in which case it is not subject to decay. However, there are some places (such as outside steps, railings, fence posts, porch columns, and exterior paneled doors) where joints in the structure may admit enough water to permit stain or decay to develop in nondurable kinds of wood. In such cases, the performance of both wood and paint may often be improved by treating the wood with preservatives or water-repellent preservatives before painting. Treatment of the wood with water-repellent preservatives at least several days before erection has been found effective in minimizing such penetration of water.

Wood Floors

Wood floors exposed to the weather (such as porch floors) are commonly painted with porch and deck paint made for the purpose. Generally, two coats are needed. Natural finishes are rarely considered durable enough for floors exposed to weather. Concrete floors may be painted, if properly treated.

Interior floors made of hardwoods most often are given a clear finish, though new wood can be stained and coated. Shellac used to be the most common finish, and when it is used, three coats are applied and the surface is gone over lightly with steel wool after the first and again after the second coat has been applied. Shellacked floors may be waxed if desired.

The most popular clear finish today is polyurethane. It dries clear and hard and does not require waxing.

Some woods need to be filled before being finished. Oak flooring should be filled with wood filler (either natural or colored) before a clear coating is applied. When floor-seal finish is used, application of filler is optional.

Preparation

The finest paint, applied with the greatest skill, will not produce a satisfactory finish unless the surface has been prepared properly. The basic principles are simple. They vary somewhat with different surfaces and, to some extent, with different paints. However, the goal is the same: to provide a surface with which the paint can make a strong, permanent bond. In general, keep the following in mind:

- Surface must be clean, smooth, and free from loose particles such as dust or old paint. Use sandpaper, a wire brush, or a scraper (Figure 5-19).

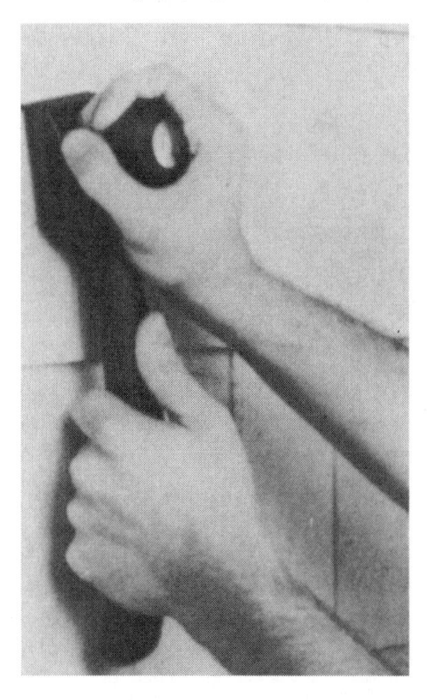

Figure 5-19 Scrape off all loose paint before painting.
(Courtesy of Stanley Tools)

- Oil and grease should be removed by wiping with mineral spirits. If a detergent is used, it should be followed by a thorough rinse with clean water.
- Chipped or blistered paint should be removed with sandpaper, a wire brush, steel wool, or a scraper.
- Chalking or powdered paint should be removed with a stiff bristle brush or by scrubbing with water mixed with household washing soda or trisodium phosphate (TSP), which is sold in hardware stores. If the old surface is only moderately chalked and the surface is relatively firm, an oil primer can be applied without the prior use of a stiff brush. The primer rebinds the loose particles and provides a solid base for the paint.
- Loose, cracked, or shrunken putty or caulk should be removed by scraping.

- If new glazing compound, caulking compounds, and sealants are used, they should be applied to a clean surface and allowed to harden before paint is applied. If the caulk is a latex type, latex paint can be applied over it immediately without waiting for the caulk to harden.
- Damp surfaces must be allowed to dry before paint is applied, unless you are using a latex paint.

Preparing Wood

Follow these steps when preparing wood:

1. Scrape clean all areas where sap (resin) has surfaced on the wood, and sand smooth prior to application of "knot sealer." Small, dry knots should also be scraped and thoroughly cleaned, then given a thin coat of sealer before applying wood primer.
2. Fill cracks, joints, crevices, and nail holes with glazing compounds, putty, or plastic wood and sand lightly until the spots are flush with the wood. Always sand in the direction of the grain—never across it.
3. Sand smooth new wood surfaces to be stain-finished. Open grain (porous) wood should be given a coat of paste filler before the stain is applied (paste fillers come in various matching wood colors). The surface should be resanded. Read manufacturer's instructions carefully before applying paste fillers.

Preparing Masonry

Surfaces such as plaster, gypsum, cement, and drywall should be dry and clean. If the surface is cracked, sand it smooth and then fill with spackling compound or some other recommended crack filler. After the repaired surface is dry, sand lightly until smooth, then wipe clean.

Follow these steps when preparing masonry:

1. Allow new plaster to dry for 30 days before painting.
2. Roughen unpainted concrete and stucco with a wire brush to permit a good bond between the surface and the paint.
3. Be sure to wash new concrete surfaces with detergent and water. This should remove any film left over from oil or from the compound used for hardening the concrete during the curing process.
4. Remove *efflorescence* (the crystalline deposit that appears on the mortar between the bricks in a brick wall) by using

undiluted vinegar or a 5 percent muriatic-acid solution. After scrubbing with acid, rinse the surface thoroughly.

Caution: When using muriatic acid, wear goggles and gloves for protection.

Preparing Metal

Follow these steps when preparing metal:

1. Clean new metal surfaces such as galvanized steel, aluminum, or tin with a solvent such as mineral spirits to remove the oil and grease applied to the metal as a preservative by manufacturers.
2. Remove rusted or corroded spots by wire-brushing or with coarse sandpaper. Chemical rust removers are also available from paint and hardware stores. Paint will not adhere well when applied over rusted or corroded surfaces.
3. Allow galvanized steel (such as that used for roof gutters) to weather for about six months before painting. If earlier painting is necessary, wash the surface with mineral spirits or Varnish Makers & Painters (VM & P) naphtha, and then apply a primer recommended specifically for galvanized surfaces.

Getting Ready

Before you brush, roll, or spray, there are certain preparations you should make to ensure a good job with a minimum of effort, errors, and spattering. The following precautions may seem obvious, but they are often overlooked:

- *Protect other surfaces*—Cover floors and furnishings with dropcloths. You can use tarps, old sheets, or the inexpensive plastic sheets designed for the purpose (Figure 5-20).
- *Clean up as you paint*—Wet paint is easy to remove; dry paint is hard to remove. Use turpentine or other thinner to remove oil paint, water to remove latex. If paint is dropped on an asphalt tile floor, do not attempt to remove it with mineral spirits or turpentine because this may permanently damage the tile. If the paint will not come off with a damp cloth, let it dry and then scrape it off.
- *Rub protective cream or Vaseline onto your hands and arms*—A film of this cream will make it easier to remove paint from your skin when the job is done. Old gloves or throwaway plastic gloves and aprons are also useful.

Figure 5-20 Moving furniture to room center and covering it up as shown is a good idea for an interior job.

- *Check the condition of the paint*—When you buy new paint of good quality from a reputable store, it is usually in excellent condition. Nevertheless, after stirring the paint thoroughly, you should examine it for lumps or color separation. Do not use the paint if there are any signs of these or the following conditions:
  - On removal of the container lid, old paints release a foul odor (especially latex paints) or show signs of lumps or curdling. These paints are probably spoiled and should be discarded.
  - If there is a skin on the surface of the paint when you open the container, remove as much of the hardened film as possible with a scraper or knife and strain the paint through a cheesecloth or fine wire mesh such as window screening. If you fail to do this, bits of the skin will show up with

exasperating frequency to spoil the appearance of your paint job.

- If the settled layer should prove to be hard or rubbery, and resists stirring, the paint is probably too old and should be discarded.

New paints are usually ready for use when purchased and require no thinning unless they are to be applied with a sprayer. Get the advice of the dealer when you buy the paint, and check the label before you mix or stir. Some manufacturers do not recommend mixing as it may introduce air bubbles.

If mixing is required, it can be done at the paint store by placing the can in a mechanical agitator, or you can do it at home with a mixing stick.

If you open the can and find that the pigment has settled, use a clean paddle and gradually work the pigment up from the bottom of the can, using a circular stirring motion. Continue until the pigment is evenly distributed, with no signs of color separation.

- *Protect the paint between jobs*—Between jobs, even if it is only overnight, cover the paint container tightly to prevent evaporation and thickening, and to protect it from dust. Oil-base and alkyd paints may develop a skin from exposure to the air.

  When you finish painting, clean the rim of the paint can thoroughly and put the lid on tight. To ensure that the lid is airtight, cover the rim with a cloth or piece of plastic film (to prevent spattering) and then tap the lid firmly into place with a hammer.

- *Priming*—Previously painted surfaces usually do not require primer coats except where the old paint is worn through or the surface has been damaged.

Wood Surfaces

For wood surfaces, keep the following in mind:

- Unpainted wood to be finished with enamel or oil-base paint should be primed with enamel undercoat to seal the wood and provide a better surface. If the unpainted wood is not primed, the enamel coat may be uneven.
- Unpainted wood to be finished with latex should first be undercoated. Water-thinned paint could raise the grain of the bare wood and leave a rough surface.

If Clear Finishes Are Used

Softwoods such as pine, poplar, and gum usually require a sealer for controlling the penetration of the finished coats. In using stain, a sealer is sometimes applied first to obtain a lighter, more uniform color. Open-grain hardwoods such as oak, walnut, and mahogany require a paste wood filler, followed by a clear wood sealer. Close-grain hardwoods such as maple and birch do not require filler. The first coat may be a thinned version of the finishing varnish, shellac, or lacquer (Figure 5-21).

Figure 5-21 Polyurethane dries in an hour. It is good on unpainted wood. *(Courtesy of United Gilsonite)*

Masonry Surfaces

Smooth, unpainted masonry surfaces and plaster or plasterboard can be primed with latex paint or latex primer-sealer. The color of the first coat should be similar to the finish coat. Coarse, rough or porous masonry surfaces (such as cement block, cinder block, and concrete) cannot be filled and covered satisfactorily with regular paints. Block filler should be used as a first coat to obtain a smooth sealed surface over which almost any paint can be used. Keep the following in mind:

- Unpainted brick, while porous, is not as rough as cinder block and similar surfaces and can be primed with latex primer-sealer or with an exterior-type latex paint.

- Enamel undercoat should be applied over the primer where the finish coat is to be a gloss or semi-gloss enamel.
- Follow carefully the manufacturer's label instructions for painting masonry surfaces.

Metal Surfaces

For metal surfaces, keep the following in mind:

- Unpainted surfaces should be primed for protection against corrosion and to provide a base for the finish paint. Interior paints do not usually adhere well to bare metal surfaces, and provide little corrosion-resistance by themselves.
- Prime paints for bare metal surfaces must be selected according to the type of metal to be painted. Some primers are made especially for iron or steel; others are for galvanized steel, aluminum, and copper.
- An enamel undercoat should be used as a second primer if the metal surface is to be finished with enamel. That is, apply the primer first, then the undercoat, and finally the enamel finish. Most enamel undercoats need a light sanding before the topcoat is applied.

Paints for Light-Wear Areas

When working with light-wear areas, keep the following in mind:

- Latex interior paints are generally used for areas where there is little need for periodic washing and scrubbing (for example, living rooms, dining rooms, bedrooms, and closets).
- Interior flat latex paints are used for interior walls and ceilings since they cover well, are easy to apply, dry quickly, are almost odorless, and can be quickly and easily removed from applicators.
- Latex paints may be applied directly over semi-gloss and gloss enamel if the surface is first roughened with sandpaper or liquid sandpaper. If the latter is used, follow carefully the instructions on the container label.
- Flat alkyd paints are often preferred for wood, drywall, and metal surfaces because they are more resistant to damage. Also, they can be applied in thicker films to produce a uniform appearance. They wash well, better than interior latex paints. They are odorless.

Paints for Heavy-Wear Areas

When working with heavy-wear areas, keep the following in mind:

- Enamels (including latex enamels) are usually preferred for kitchen, bathroom, laundry room, and similar work areas because they withstand intensive cleaning and wear. They form especially hard films, ranging from flat to full gloss.
- Fast-drying polyurethane enamels and clear varnishes provide excellent hard, flexible finishes for wood floors. Other enamels and clear finishes can also be used. However, unless specifically recommended for floors, they may be too soft and slower drying, or too hard and brittle.
- Polyurethane and epoxy enamels are also excellent for concrete floors. For a smooth finish, rough concrete should be properly primed with an alkali-resistant primer to fill the pores. When using these enamels, adequate ventilation is essential for protection from flammable vapors.

Clear Finishes for Wood

When using clear finishes for wood, keep the following in mind:

- Varnishes form durable and attractive finishes for interior wood surfaces such as wood paneling, trim, floors, and unpainted furniture. They seal the wood, forming tough, transparent films that will withstand frequent scrubbing and hard use, and are available in flat, semi-gloss, or satin and gloss finishes.
- Most varnishes are easily scratched, and the marks are difficult to conceal without redoing the entire surface. A good paste wax applied over the finished varnish (especially on wood furniture) will provide some protection against scratches.
- Polyurethane and epoxy varnishes are notable for durability and high resistance to stains, abrasions, acids, alkalis, alcohol, and chemicals. Adequate ventilation should be provided as protection from flammable vapors when these varnishes are being applied.
- Shellac and lacquer have uses similar to most varnishes, and these finishes are easy to repair or recoat. They apply easily, dry fast, and are useful as a sealer and clear finish under varnish for wood surfaces. The first coat should be thinned as recommended on the container, then sanded very lightly and

finished with one or more undiluted coats. Two coats will give a sheen, and three a high gloss.

Painting Exterior Surfaces

The durability of an exterior paint job depends greatly on the following:

- Surface preparation
- The quality of paint selected
- The skill of application
- The proper spacing of repaintings
- The protection of surfaces from the sun and rain
- Climatic and local weather conditions

Before you paint exteriors, check the weather. You can easily ruin a paint job if you forget to consider the weather. Excessive humidity or extremely cold weather can cause problems. Keep the following in mind:

- Unless you are using latex paint, you should not paint on damp days. Moisture on the painting surface may prevent a good bond.
- If humidity is high, check the surface before painting. If you can feel a film of moisture on the surface, it would be better to wait for a dryer day.
- Exterior painting is not recommended if the temperature is below 50°F or above 95°F because you may not be able to get a good bond. This is especially critical if you are using latex paint.

Allow more drying time in damp or humid weather. The label on the can will tell you the normal drying time, but test each coat by touch before you add another. When paint is thoroughly dry, it is firm to the touch and is not sticky.

Latex paints can be used even if the surface is not bone dry. The best time for exterior painting is after the morning dew has evaporated.

Before you start on the job, make a thorough inspection tour and check the surface condition of window and doorframes and surrounding areas, bases of columns of porches and entranceways, steps, siding, downspouts, under-eave areas, and anywhere that moisture is likely to collect.

Priming Wood

When priming wood, keep the following in mind:

- The tendency of wood to expand and contract during changes in temperature and humidity makes it imperative that a good wood primer be applied to provide the necessary anchorage for the finish paint.
- Surfaces such as wood siding, porches, trim, shutters, sash doors, and windowsills should be primed with an exterior primer intended for wood. Application should be by brush; surfaces should be thoroughly dry.
- Painted wood usually does not need priming unless the old paint has cracked, blistered, or peeled. Defective paint must be removed by scraping or wire brushing (preferably down to the bare wood) and then primed.
- Scratches, dents, recesses, and raw edges should be smoothed and then touched up with a suitable exterior primer.

Priming Masonry

When priming masonry, keep the following in mind:

- New masonry surfaces should be primed with an exterior latex paint, preferably one specifically made for masonry.
- Common brick is sometimes sealed with a penetrating type of clear exterior varnish to control efflorescence and *spalling* (flaking or chipping of the brick). This varnish withstands weather, yet allows the natural appearance of the surface to show through.
- Coarse, rough, and porous surfaces should be covered with a fill coat (block filler), applied by brush to thoroughly penetrate and fill the pores.
- Old painted surfaces that have become a little chalky should be painted with an exterior oil primer to rebind the chalk. If there is much chalk, it should be removed with a stiff brush or by washing with household washing soda or TSP mixed with water.

Priming Metal

When priming metal, keep the following in mind:

- Copper should be cleaned with a phosphoric acid cleaner, buffed and polished until bright, and then coated before it discolors. Copper gutters and downspouts do not require painting. The protective oxide that forms on the copper surface darkens it or turns it green, but does not shorten the life of the metal. Copper is often painted to prevent staining of adjacent painted surfaces.

- Zinc chromate type primers are effective on copper, aluminum, and steel surfaces, but other types are also available for use on metal.
- Galvanized steel surfaces (such as gutters and downspouts) should be primed with recommended special primers, since conventional primers usually do not adhere well to this type of metal. A zinc-dust zinc-oxide primer works well on galvanized steel. Exterior latex paints (but not oil paints) are sometimes used directly over galvanized surfaces.
- Unpainted iron and steel surfaces will rust when exposed to the weather. Rust, dirt, oils, and old loose paint should be removed from these surfaces by wire brushing or power-tool cleaning. The surface should then be treated with an anticorrosive primer.

Finishing

When finishing, keep the following in mind:

- All exterior surfaces, properly primed or previously painted, can be finished with either exterior oil paint or exterior latex paint.
- Latex paints are easy to apply, have good color retention, and can be used on slightly damp surfaces.
- Oil- or alkyd-base paints have excellent penetrating properties. They provide good adhesion, durability, and resistance to abrasion and blistering on wood and other porous surfaces.
- Mildew, fungus, and mold growths on exterior surfaces are a problem in areas where high temperature and humidity are prevalent. Use paint that contains agents to resist bacterial and mold growth.
- Colored exterior house paints must resist chalking so that colors will not fade and erosion of the paint film will be minimized. The manufacturer's label will indicate whether the paint is a nonchalking type. Some white exterior house paints are expected to chalk slightly as a means of self-cleaning.

Paint Failures

Understanding why paint fails will help you avoid paint problems. There are a number of reasons for exterior-paint failures. One of the major causes of paint failure is moisture. Quality of paint and the method of application are others.

Following are some common things to keep in mind about paint failures:

- If steel nails have been used for the application of the siding, disfiguring rust spots may occur at the nail heads, particularly where they are exposed. Spotting is somewhat less apparent where steel nails have been set and puttied. Similarly, the spotting may be minimized, in the case of flush nailing, by setting nail heads below the surface and puttying. The puttying should be preceded by a priming coat.
- Brick and other types of masonry are not always waterproof, and continued rains may result in a damp interior wall or wet spots where water has worked through. If this trouble persists, it may be a good idea to use a waterproof coating on the exposed surface. Transparent coatings can be obtained for this purpose.
- Caulking is usually required where a change in material occurs (such as that of wood siding abutting against brick chimneys or walls). The wood should always have a prime coating of paint for proper adhesion of the caulking compound. Caulking guns with cartridges can be obtained and are the best means of waterproofing these joints.
- Rainwater flowing down over wood siding may work through butts and end joints and sometimes may work up under the butt edge by capillary action. Setting the butt end joints in white lead is an old-time custom that is very effective in preventing water from entering. Painting under the butt edges at the lap adds mechanical resistance to water ingress. Moisture changes in the siding cause some swelling and shrinking that may break the paint film. Treating the siding with a water repellent before it is applied is an effective method of reducing capillary action. For houses already built, the water repellent could be applied under the butt edges of bevel siding or along the joints of drop siding and at all vertical joints. Excess repellent on the face of painted surfaces should be wiped off.
- Two of the most common paint problems are peeling and bleeding. Peeling usually has to do with moisture, though

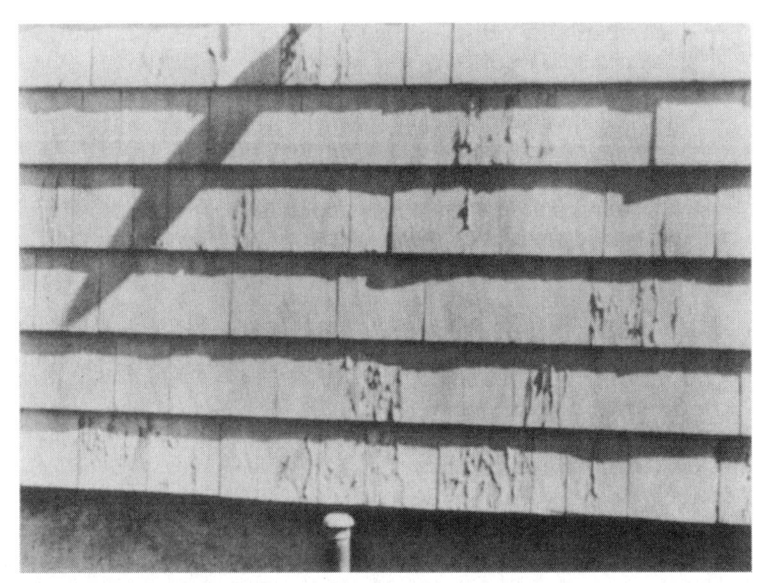

Figure 5-22 Peeling.

Figure 5-23 Bleeding.

incompatibility of paints also causes it (Figure 5-22). At any rate, check for moisture penetration, and check paint compatibility. All peeling paint must come off before painting. Bleeding is also sometimes caused by paint incompatibility or, on interiors, it can occur if you apply flat paint over plaster (Figure 5-23). It can also occur when the paint used is too thin, or when flat paint is applied over enamel. Check these situations out before painting.

Summary

Painting appears to be a simple operation, but it takes concentration and the use of proper tools and techniques to do a good job.

You can normally use either a brush or roller, though in some instances (such as when painting a highly textured surface), only a brush will serve well. Take care and pick quality brushes and rollers and take good care of the tools. Clean them well after use and store properly.

A paint sprayer may also be used, but the paint must be of the proper consistency. Make sure surrounding areas are covered. Before using any kinds of paint in any application, read the label and follow the manufacturer's directions.

If two or more coats are required, the surfaces are first prepared and primed. This is as important (even more important) when applying the finish coat.

Working with color is also an important aspect of painting. Colors may be mixed from scratch (white with color added), store-bought, or bought already tinted.

Problems are also a part of painting, and water is the chief culprit. All potential problems must be cleared up before painting begins.

Review Questions

1. What are the common tools used for painting?
2. What are typical preparation jobs?
3. What is the chief problem when paint problems occur?
4. When is spray-painting a good idea?
5. What is creating color from "scratch"?
6. What is a tint?
7. What is a shade?
8. What are the three primary colors?
9. How is paint colored to fit an existing shade or color?
10. Where do you get trisodium phosphate? Why do painters use it?

Chapter 6

Design and Layout of a Workshop

Refined woodworking skills are a natural extension of carpentry skills. This chapter is for the professional carpenter who is ready to set up a workshop as a place to develop and refine those skills.

Often a workshop is set up on the construction site, right inside the house being built. However, there are advantages to having a dedicated workshop:

- *Economy*—Buy materials in bulk at lower cost and store for later use. Storage space is always at a premium on a construction site, where materials must be quickly bought and installed.
- *Schedule*—Production of architectural woodwork can begin before closing in at the site. Work can continue at the workshop during inclement weather, or when the site becomes crowded with several trades working at once.
- *Efficiency*—Multiple machines can be left set up to do specific operations. This saves setup time and results in a higher rate of production than when resetting just one machine for several different operations.

As anyone in the building trades knows, there is more than one way to do any job. Designing a workshop is no different. The shop illustrated in this chapter is a real working shop. This shop is used as an example, but it is by no means the ideal shop. In fact, the owner began a description of his shop with a list of ways the shop could be improved. Yet, the shop does show good layout and design that matches the people who work there and the work they do.

Rather than just copy a design, think about the work to be done and the resources available to design a shop to meet your own specific needs.

The Design Process

Begin by thinking about the kind of work to be done in the shop. Develop a business plan to ensure that there is enough money, as well as other resources, to support the shop. Lay your thoughts out on paper. Then consider what specific features the shop needs. A basic strategy is to keep the design flexible.

Consider the Type of Work

A shop that only produces large runs of specialty flooring would be very different from one that produces custom cabinetwork and fine furniture. Markets that support the shop will change in the future. Changing work needs flexibility in layout. Being able to adapt quickly is an important marketing advantage the small shop has over larger millworks.

Architectural woodworking is needed when the residential market is up. Cabinetmaking and furniture can be done only when workers with these specialized skills are available. Expanding to include all kinds of woodworking may be necessary when the economy declines and specialty work is not available.

If most of the production is architectural millwork for high-end residential housing, it would require the ability to work with special-sized trim boards that may have custom molded edges, paneling for walls, flooring, and some special cabinetwork (including kitchens).

Develop a Business Plan

Building and operating a workshop is an expensive proposition. A business plan demonstrates that the expense will be balanced with income to support it. The plan shows in dollar terms where the money comes from and where it will be spent. It shows whether the shop will make or lose money over time.

Adding up the resources available for building and equipping a shop is an important beginning. On paper, write down a list of everything you have that could contribute to the shop. Include money, equipment, and space that may already be available. Do not overlook or underestimate the time it will take to design the shop and put it together.

To support a shop financially, a market with a big profit margin is needed. This might be high-end custom residential work or commercial fittings for offices and retail stores. Whatever the market is, you must know it will last long enough to support the shop well into the future.

Typically, the costs of a shop are broken down into two parts. The first is a one-time major capital investment. This would cover the cost of putting up a building, or, if existing space is available, the cost of initial machinery.

The second cost is that of continuing overhead operations. Heat, electricity, and supplies are included here. Maintenance and repairs of both machinery and building and grounds should be included.

Both capital and overhead costs are recovered by adding a margin to the price that is charged for work done in the shop.

Starting with a small shop and one or two workers to support it minimizes the capital investment needed. It usually takes the labor of three or more workers to ensure a return on the investment it takes to capitalize a fully equipped production workshop.

Size the Shop

No matter how much space they have, most shop owners say they need more. Enclose as much space as you can comfortably afford, but do not go overboard. If the shop is successful, it can be expanded later. The cost of future expansion can be minimized if planned for in the beginning.

Consider how many workers will use the shop, space for machines, and how much storage will be needed.

Lay Out a Plan

To determine the overall size needed, make a scale drawing of the proposed floor plan. Then cut pieces of paper or cardboard to represent each machine to scale. It's much easier to shuffle paper than 900-pound machines as you try different layouts.

Machinery Layout and Work Flow

To begin positioning machinery, think about how work will flow through the shop (Figure 6-1). Arrange machinery to reduce the number of times materials and parts are handled as they pass through the shop. Following are typical operations of work in steps from rough lumber to finished architectural and cabinetwork:

- Cut-off to length
- Rip to width

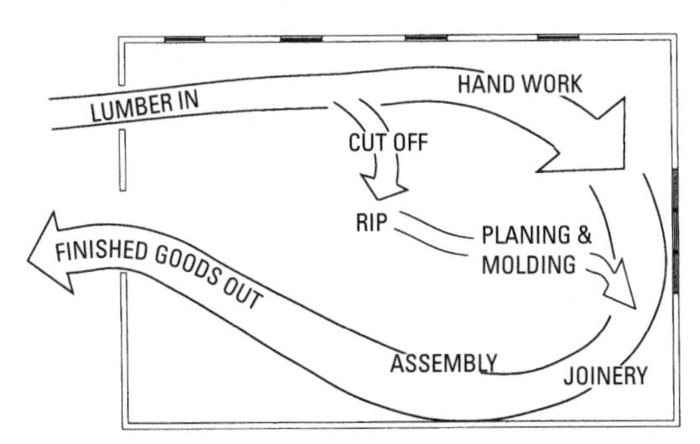

Figure 6-1 Material movement through the shop should flow smoothly to reduce handling time between operations. Here lumber is moved in through the small door and cut off to length at the radial arm saw. Then it is ripped to width at the table saw. After planing and molding, joints are cut and parts are moved to the assembly area. After final assembly, finished goods are shipped out.

- Thickness planing
- Molding
- Handwork
- Sanding
- Joinery
- Assembly

Use this list to build a project in your imagination. Then, as you work through a typical project, the lumber is cut up into smaller and smaller parts until final assembly of, for example, cabinets, when you end up with a few very large assemblies. If you plan to build kitchens, be sure there is enough space to store all the completed cabinets. Try different machinery arrangements and shop sizes to determine your particular needs.

Clearance Paths

Each machine claims floor space beyond its own "footprint." This is the clearance path boards and parts need as they pass through the machine. Locate each machine so the path is clear and free of nearby machinery, walls, or posts (Figure 6-2). Clearance paths of different machines can overlap if the machines will not be used at the same time on a consistent basis.

Make sure there is enough space for the workers as well. If there are more than a few workers, leave plenty of room for the passage of carts or dollies. They will be carrying stacks of lumber or parts. This is especially important in a high-production setting. That is where there are many workers. They may not be aware of what is happening around them.

Loading and Unloading

Make unloading lumber into the shop and loading up finished goods as easy as possible with large doors and a loading dock. You may want to combine both with double doors in the end of the shop that has a sill at the loading dock with its height above the ground just outside the doors (Figure 6-3).

Multiple Machines

A production shop gains its main economic advantage by having several machines setup and standing ready to produce goods. Primary machines are kept free to do the basics, while secondary support machines of the same type are left setup to do special operations.

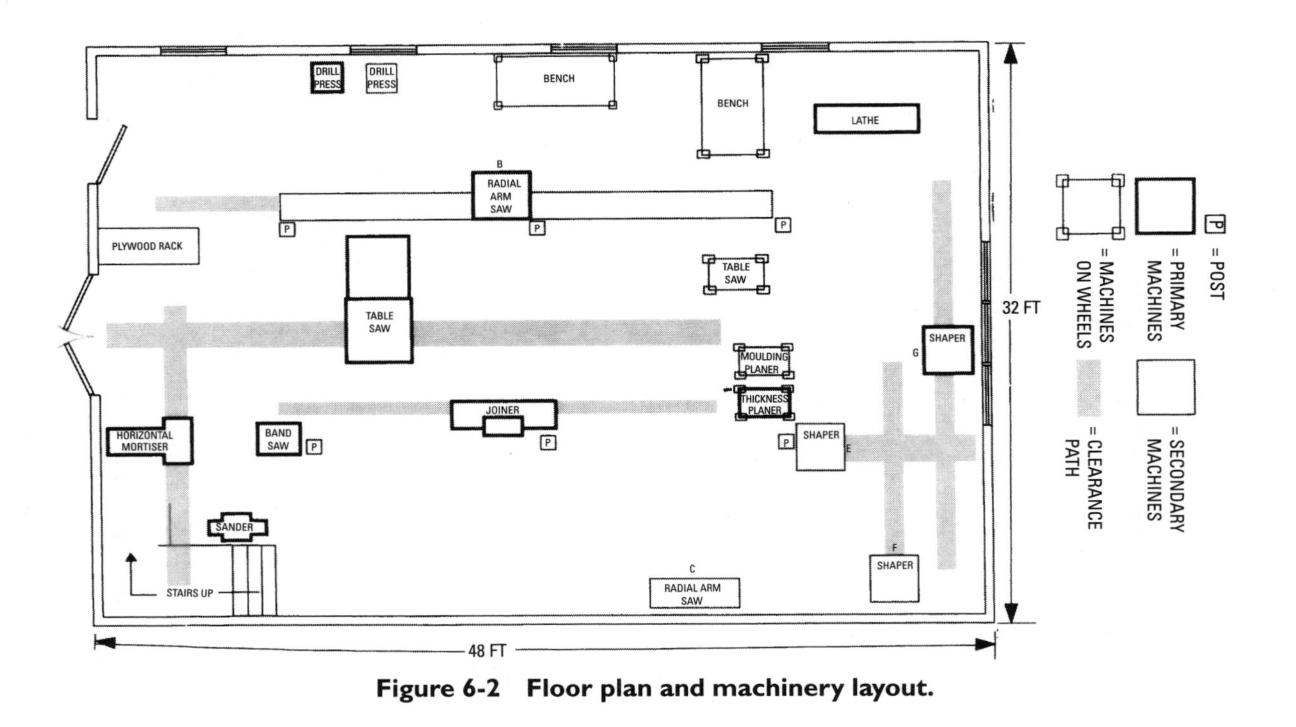

Figure 6-2 Floor plan and machinery layout.

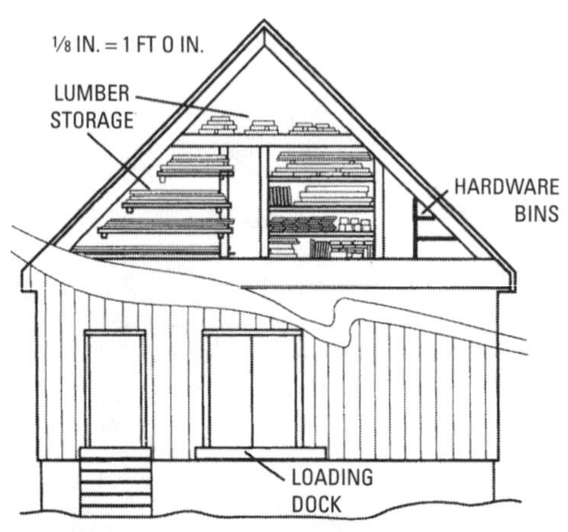

Figure 6-3 End view of shop storage areas.

Consider the example of making a framed panel door. A radial arm saw and table saw are used to cut out the stiles, rails, and panels. A radial saw is always available for ordinary cutoffs. A table saw is ready for general ripping anytime.

Secondary machines are used in most operations to complete parts for an example door.

- A secondary *radial arm saw* for cutting tenon shoulders is needed.
- A secondary *table saw* comes in handy for cutting tenon cheeks.
- A secondary *shaper* is needed to raise the edge of panels.
- A secondary *shaper* can be used to cope moldings on the rails.

The primary shaper can be used to mold or "stick" the stiles.

The special setups of these secondary machines are left in place. This means a door can be made quickly, leaving the primary machines free for other, more general work. When machines are used together like this, locate them close together on the layout. An example is the cluster of three shapers in the lower right of Figure 6-2.

Features to Consider

Once you have the basic operations and machinery layout complete, consider how to handle the following features. Change the basic layout as needed to accommodate the features required.

Flooring

Concrete floors can cause fatigue in legs and feet when working long hours. The hard floors even cause serious health problems over the long term. Wood flooring provides a little give. This slight give eases strain.

If you happen to drop a wooded part or cabinet, it is much less likely to be damaged by a wooden floor. Jigs and fixtures can be conveniently screwed to a wood floor as needed.

Add a wood floor over a concrete floor by laying 2-inch × 2-inch or 2-inch × 4-inch framing lumber as sleepers 16 inches on-center. Then lay softwood tongue and groove flooring onto the sleepers nailing every 12 inches. Remember, this will reduce the ceiling height.

Heat

Fire safety should be a primary consideration in determining the type of heating system. The low distribution temperatures of forced hot water with a heating plant in a small building separate from the shop may be the safest. Forced hot-air distribution will warm up the shop quickest on cold mornings. Burning waste wood for fuel makes sense only if it is done carefully and safely.

Benches, Dollies, and Carts

Sturdy workbenches are especially important if handwork will be a major part of the production process. When sawing, planing, or carving, any movement in the bench reduces efficiency significantly. It can even affect quality. Massive bench tops at least 3 inches thick help reduce vibration. Crossbracing below reduces movement.

Make at least one bench in the shop movable by adding wheels with brakes so it can be locked in position.

Wheeled tables, carts, or dollies are used for stacking and efficiently moving large numbers of parts between operations.

Lighting

The ideal is to let in as much natural light as possible, and then supplement it with fluorescent fixtures to provide low-cost light with little glare or shadows. A simple test for shadows is to stand a 1-inch dowel on each machine table and bench. If a shadow is cast in any direction, add to or adjust the position of lights to eliminate the shadow.

Paint ceilings and walls white to get the full use of available light.

Material Storage

Keep the production floor free for work by storing everything not currently needed. This may mean another room or a building nearby.

A better solution is to think in three dimensions and store lumber and other supplies in a loft or second story of the shop building (Figure 6-3).

Plywood storage on the production floor should be in a rack that holds the sheets vertically on edge to reduce the necessary floor space. Even heavy plywood sheets can be sorted and selected easily when stored on edge. Long-term storage of large quantities of plywood should be horizontal on a flat surface to prevent warping. Put blocks under plywood to keep it off concrete floors to prevent moisture damage.

Waste Storage and Handling

As lumber is drawn from stock and used for production, waste is generated. The waste may range from sawdust with little or no value to a 12-foot plank of expensive walnut hardwood that had only 2 feet used off one end. In between are a variety of sizes and shapes that may be more or less useful sometime in the future.

Clearly, the walnut should be saved for future use and the sawdust should be gotten rid of as soon as it is generated. Nevertheless, should what is in between be saved or tossed? If you are in business to make money, the decision is an economic one. Get rid of waste when it costs more to store and handle than it does to buy new wood.

Often in the small shop, waste is saved at a far greater cost than just getting rid of it. Costs include the labor of handling it several times and the value of the space where it is stored.

The line between toss or save is drawn differently in every shop. Consider the dollar values of the wood, labor, space, and the kind of products made. Then set a policy to deal with it. The policy might be to toss everything less than 2 feet long, or to save all hardwoods but toss all softwoods. Whatever the policy is, stick to it. After a year, inventory the waste stock and evaluate the policy to see if it is making or losing money.

Set up waste storage that is space-efficient and easy to use. Waste stored where it can't be seen will soon be forgotten and never used.

Summary

A dedicated workshop can save money, time, and effort. Like any project, setting up a workshop requires careful planning. A flexible design must be developed that considers the type of work to be done in the shop and the resources available. Lay out the design on paper to help visualize it. Locate machinery and storage so work flows through the shop smoothly. Machines with special setups make the shop more productive. Carefully selected features such as wood flooring, heat, benches, good lighting, and storage systems make the shop efficient.

Review Questions

1. What are the advantages to having a dedicated workshop?
2. Where do you begin the design process for a workshop?
3. What do you consider in making a workshop?
4. How big a shop do you need?
5. What kind of machinery do you need for a one-person operation?
6. What is meant by the term "workflow"?
7. How do you go about moving your proposed machinery on your floor plan for the shop?
8. Once you have the basic operations and machinery layout complete, what are the other factors you need to consider?
9. What is a good philosophy about waste storage and handling?
10. What can a dedicated workshop do for you?

Chapter 7

Equipment and Machinery

The quality of work performed by a carpenter or builder often depends on the quality of equipment and machinery used. This chapter examines the essentials that carpenters and builders might use on a variety of projects, including the following:

- Table saw
- Band saw
- Jigsaw
- Wood lathe
- Thickness planer
- Jointer
- Shaper
- Radial arm saw
- Sander
- Drill press

Table Saw

The *table saw* is one of the most popular tools in any woodworking shop or plant. Plants usually have one or more power saws used exclusively for ripping and one or more for crosscutting. The table saw can do both. Beveling and mitering can also be done with a table saw, while grooving and dadoing can also be done by means of special cutters.

In use, the stock is fed along the table and across the saw, from which the revolving blade just projects (Figure 7-1). If desired, you can tilt the blade for making angled or beveled cuts, and the tool can also be equipped with accessories such as the dado head, with which you can make rabbet cuts and grooves.

Construction

The table saw (Figure 7-2) generally consists of an *iron base* (or *frame*) on which a table is mounted, and an *arbor* (or *shaft*) that carries the saw blade or other cutter. The arbor or shaft revolves in two bearings, which are bolted to the frame. It is driven by a belt that passes over a pulley. This pulley is fastened to the shaft between two bearings.

When wood is cut on a table saw, it must be firmly held against a metal guide or fence, which can be set at any convenient distance

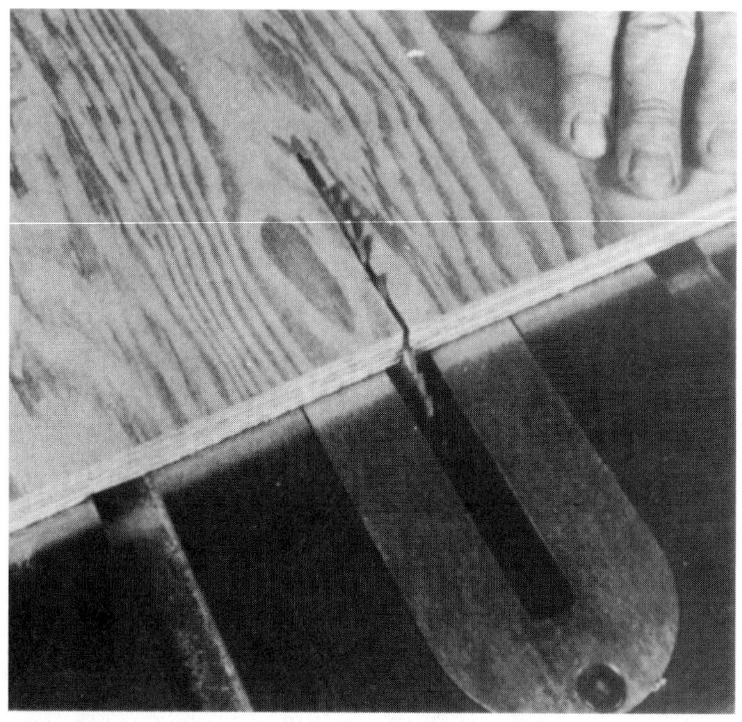

Figure 7-1 When cutting, the table saw blade should barely project from the stock. *(Courtesy of the American Plywood Association)*

from the saw blade. When the fence is used as a guide for cutting boards lengthwise, the operation is known as *ripping*.

Table saws are equipped with slots or grooves to accommodate a miter gage, which is used as a guide when sawing across a board. This operation is known as *crosscutting*.

Ripping Operation

One of the most useful operations of the table saw is that of ripping stock to its required width (Figs. 7-3 through 7-6). This operation is generally accomplished as follows:

1. The fence is set to the graduated scale at the front of the table to cut the required width.
2. The saw is adjusted (raised or lowered) to project approximately $^{1}/_{4}$ inch above the stock to be ripped.

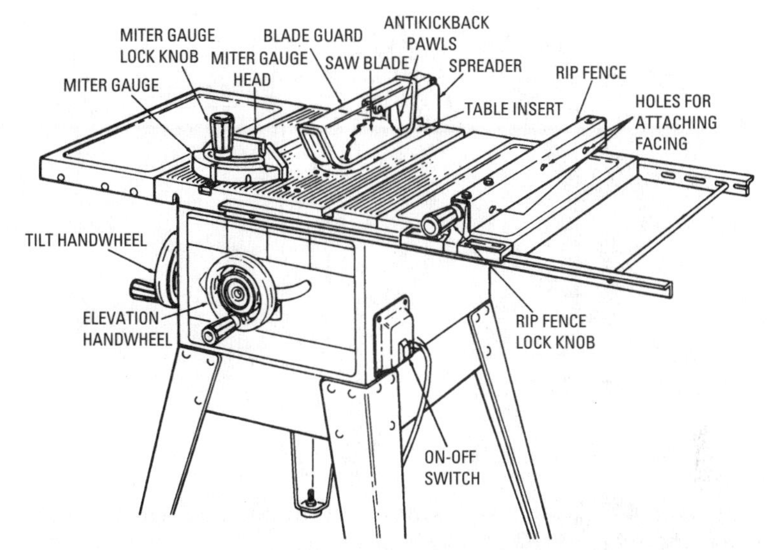

Figure 7-2 Parts of a table saw, also known as a bench saw.

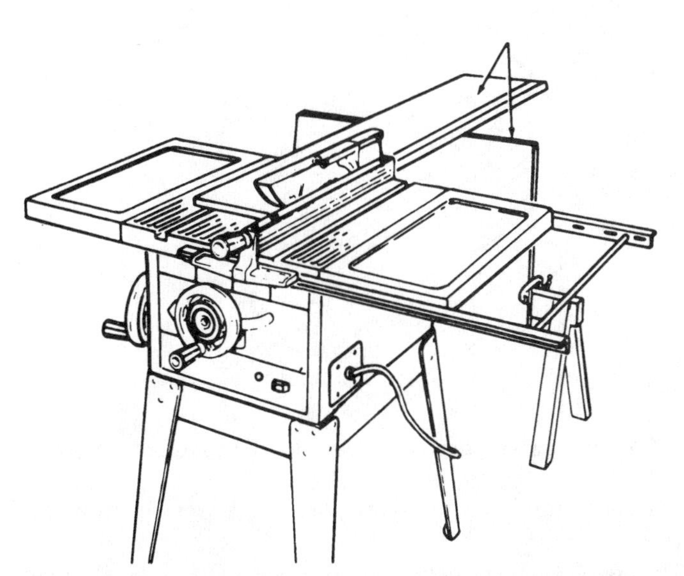

Figure 7-3 It's best to support long stock before ripping.

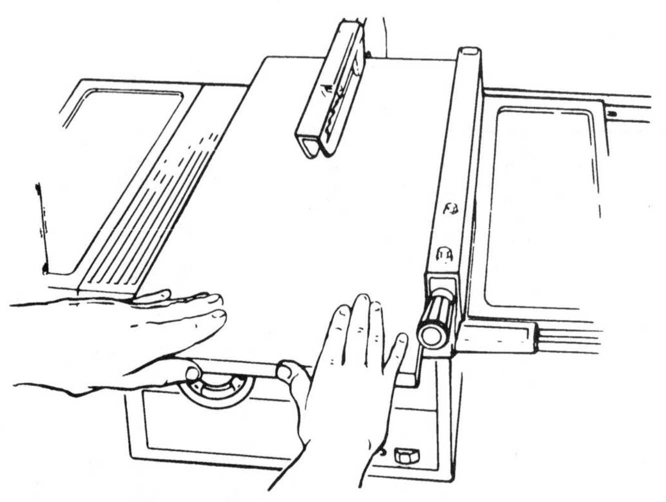

Figure 7-4 Ripping wide stock on the saw. The fence is set so that the edge of the stock rides against it.

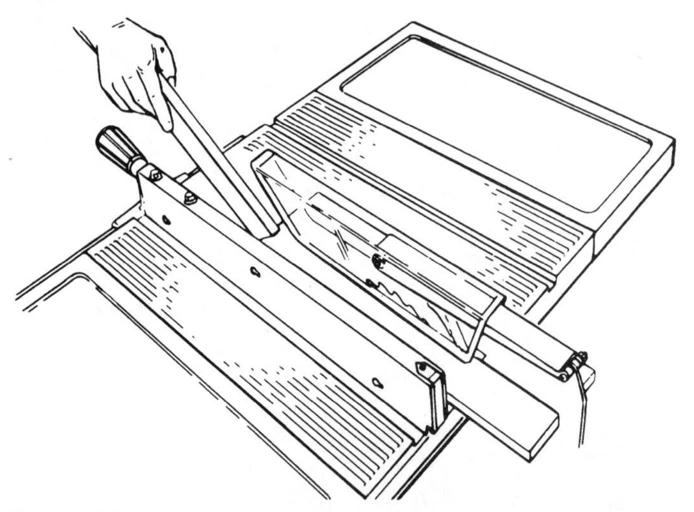

Figure 7-5 For safety, use a push stick when stock is narrow.

3. The splitter and saw guard are positioned to ensure safe operation.
4. The operation is started by holding the work close to the fence and pushing it toward the rotating saw blade with a firm, even motion. A smooth, uniform speed of feed should be used.

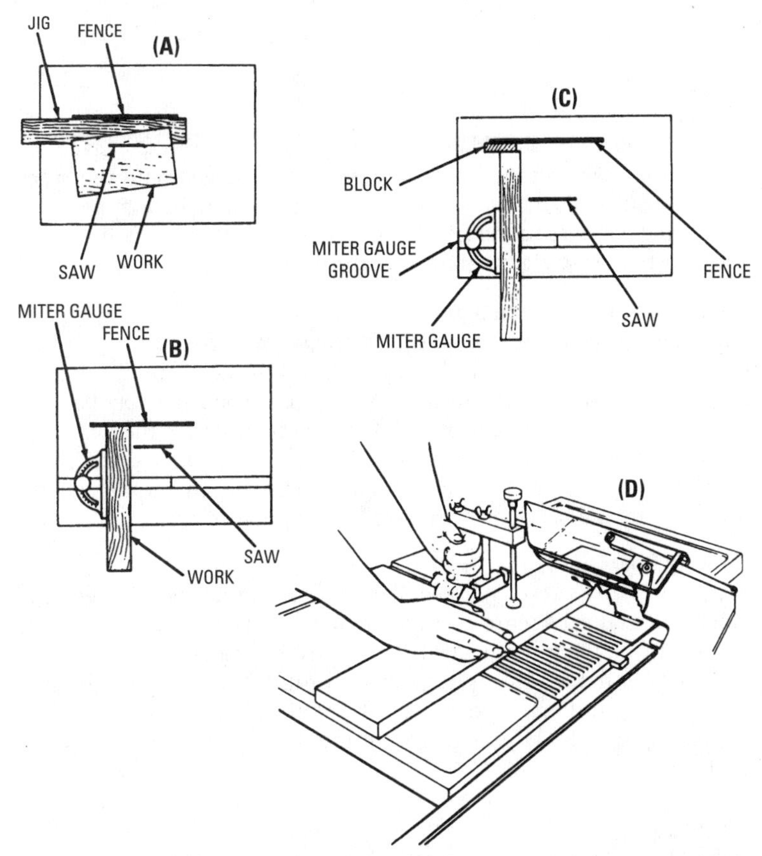

Figure 7-6 Typical sawing operations on the table saw. In taper ripping (A), work that is to be ripped on a taper cannot be guided against the fence but must be held in a tapering jig. The same idea can be applied to a number of other forms. When cutting shoulders, the stock should first be squared on one end and cut to the desired length. The miter gage is used in conjunction with the fence to bring the shoulder cut the correct distance from the end (B). When crosscutting wide pieces to length, one end should first be squared and the gage should be adjusted to the required length. Narrow pieces may be sawed to length by placing a block against the fence (C). The distance from the saw blade to the block is then the required length. Note the hold-down device (D) when making a crosscut bevel cut.

Avoid jerky movements and jamming the work through too quickly.

Caution

The operator should not stand directly behind the saw blade but should take a position slightly to either side and hold the stock near its end so that one hand will pass to the right and the other hand will pass to the left of the saw blade. A short stick is used to push narrow pieces of stock through.

Crosscutting Operation

Square crosscutting work on the table saw is performed by placing the work against the miter gage and then advancing both the gage and the work toward the rotating saw blade. The gage may be used in either table groove, although most operators prefer the left-hand groove for average work. It is essential that the miter gage be set correctly to obtain a square cut. Therefore, it is customary to test the work by means of a try square before proceeding.

Mitering Operations

Most miters are cut to an angle of 45°, because four pieces cut at this angle will make a square or a rectangle when assembled. Miters are cut by setting the miter gage at the required number of degrees. The angle of the miter for any regular polygon is obtained by dividing 180° by the number of sides and subtracting the quotient from 90°. For example, to find the angle for the miter of a pentagon (five sides), we have

$$90 - (^{180}/_{5}) = 54^\circ$$

Similarly, the angle for the miter of an octagon (eight sides) will be

$$90 - (^{180}/_{8}) = 67.5^\circ$$

The miter gage may be used in either of the table grooves and may be set on either side of the center position. When great accuracy is required on miter cuts, special miter-clamp attachments are used.

Grooving Operations

These operations consist of making grooves that are wider than those cut by ordinary saw blades. Grooves of varying widths are commonly cut on a table saw by employing a special attachment known as a *dado head*. This head is made up of two outside cutters and three or four inside cutters (Figure 7-7) by means of which grooves varying in width from $^{1}/_{8}$ to $^{13}/_{16}$ inch (or larger) can be

Figure 7-7 A typical dado-head assembly. A dado head, as shown, is made up of two outside blades $^1/_8$-inch thick together with one or more cutting fillers $^1/_{16}$-inch and $^1/_8$-inch thick, depending on the width of the groove desired. Grooves varying in widths up to 1 inch can be cut.

made using different combinations of cutters. The outside cutters generally have eight sections of cutting teeth and four raker teeth. The eight sections of cutting teeth are ground alternately left and right to divide the cut. Inside cutters of $^1/_8$ inch and $^1/_4$ inch are usually swage-set for clearance.

Grooves that are cut across the grain are termed *dadoes*. They are usually cut at right angles, but may also be cut at other angles. Dadoes that do not extend entirely from one side of a board to the other side are called *stopped dadoes, blind dadoes*, or *gains*. A gain may have one open end, or both ends may be closed (Figures 7-8 and 7-9).

Band Saw

Band saws are manufactured for many uses. For industrial ripping and resawing, large machines with blades from 3 to 5 inches wide are generally used. The type most adaptable to the general woodworking shop has blades from $^1/_2$- to 1-inch wide and is used particularly for cutting curved outlines.

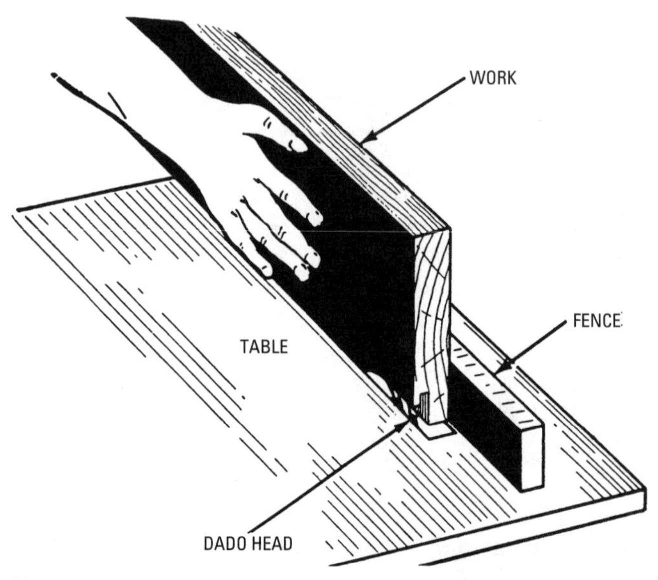

Figure 7-8 The method of cutting a rabbeted joint with a dado head. Joints of this type are used extensively in drawer construction.

Figure 7-9 The method of cutting wide grooves with a dado head.

Construction

Essentially, a band saw consists of a table, wheels, guides, saw blade, and suitable guards. The average table can be tilted, usually 45° in one direction and 10° in the other. Some, however, are built to permit bevel cutting by varying the blade angle. The band saw is generally used for resawing because of its thinner kerf.

A band saw (Figure 7-10) such as is used in woodworking shops, generally consists of an endless band of steel with saw teeth on one edge, passing over two vertical wheels and through a slot in a table. The blade is held in position by a guide.

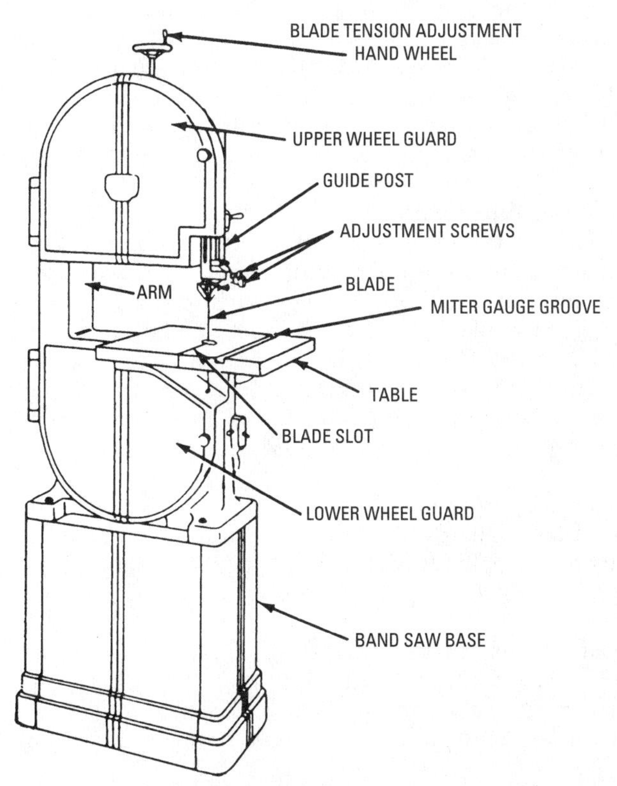

Figure 7-10 A typical band saw suitable for the small- and medium-sized woodworking shop. Tension and tilt are regulated by a convenient handwheel and a knob. The band saw gets its name from the blades used, which are continuous bands of steel.

The two vertical wheels over which the blade is fitted are usually made of cast iron. Their rims are equipped with rubber linings and are provided with adjustments for centering the saw on the rims and for giving the saw blade the proper tension under all loads.

The table supporting the work is fastened to a casting directly above the lower wheel. It is slotted for the saw blade from the center to one edge. The guide prevents the blade from twisting sideways in the slot and gives it support when cutting. Design varies for different types of saws. Some band-saw tables are equipped with a ripping fence, and some are also provided with a groove for a miter gage.

The size of the band saw depends on the diameter of the wheels, which may vary in size from 10 to approximately 40 inches. So, a saw with 10-inch diameter wheels is called a 10-inch saw. Of course, the larger the wheels, the larger, in proportion, the other parts of the saw, and consequently, the larger the size of stock that can be sawed. Other important dimensions of the band saw are the table size and the height between the table and the upper blade guide.

Straight-Cutting Operation

Although a band saw is essential for curved cutting, it may also be used for making straight cuts for both crosscutting and ripping when a table saw is not available. For all straight cuts, it is advisable to use the widest possible blade because it is easier to follow a straight line with a wide blade. The use of the miter gage for crosscutting wide stock (Figure 7-11) follows the same general procedure as that used for similar work on the table saw. In the absence of a miter gage, a wide board with square ends and sides may be used (Figure 7-12).

The use of an auxiliary wood fence fastened to the gage facilitates the handling of large boards and results in more accurate work. Unlike the auxiliary ripping fence for the table saw, the wood fence for the band saw should be kept low so that it will work under the guides.

Ripping and resawing may be performed on a band saw by using the ripping fence furnished with most saws. When the stock is worked flat on the table, the operation is ripping, and when the board is worked on edge, the operation is usually known as *resawing*.

Another type of guide frequently used in these operations is known as the *pivot block*. The pivot block consists of a specially formed wood block that is clamped to the table to hold it in the desired position. The guide is set opposite the blade and at the proper distance from it to cut the required thickness (Figure 7-13).

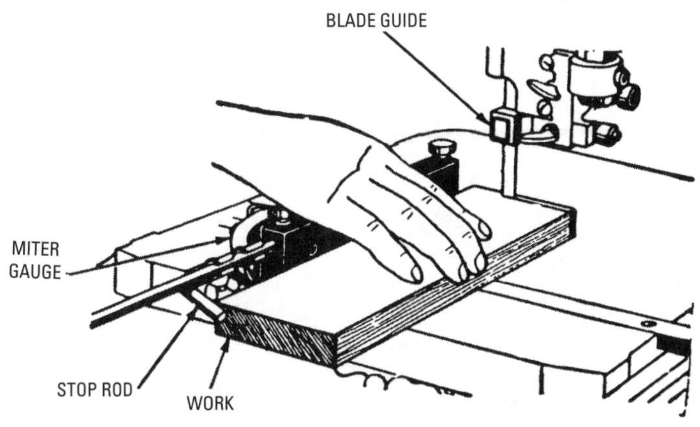

Figure 7-11 Cutting to length with a miter gage and a stop rod. The stop rod should be carried on the outer end of the miter gage.

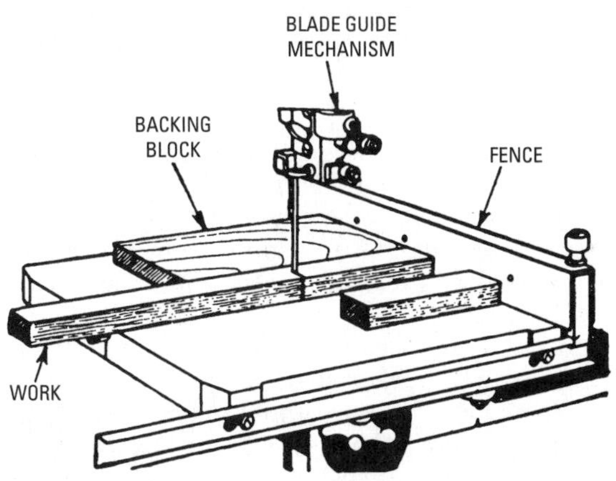

Figure 7-12 The method of cutting short pieces to length using a square board or a ripping fence when the miter gage is not available.

Cutting Arcs and Segments

When cutting arcs, the usual procedure is to first make an outline by means of a compass or divider after determining the correct radius or diameter. If several pieces all having the same curvature are to be sawed, a jig may preferably be made, in which case the arcs can be accurately cut without having to make an outline.

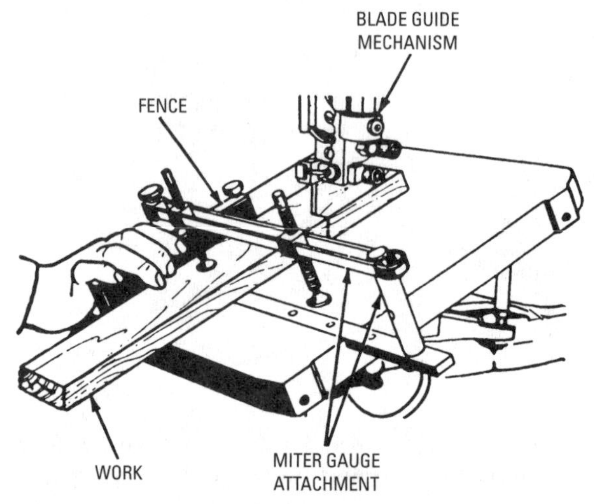

Figure 7-13 Use of the miter-gage clamp attachment. An attachment of this type is useful in many crosscutting operations and particularly when cutting at an angle with the table tilted.

Multiple-Sawing Operation

To do the best work on a band saw, it is necessary to mount it on a substantial foundation to eliminate vibration. The blades must be kept in prime condition and must be properly adjusted on the wheels. When a considerable amount of wood is to be processed, an extension to the saw table is a real convenience, if not an actual necessity.

The numerous furniture parts frequently sawed in multiple include ornate chair and table stretchers, chair bannisters, and small brackets. The more ornate items, however, are usually scrolled out on a jigsaw (discussed later in this chapter). Dependable machines are usually provided with accurate tension devices that assist the operator in securing volume production, as well as turning out high-class work.

Some jigsaws are constructed with the table and saw guides set at such an angle. The angle is with reference to the machine column so that extremely long material may be sawed easily.

When cutting very thick material, best results may be obtained by sawing one piece at a time, particularly where there are pitch spots and checks and knots to be dodged. On the other hand, anywhere from 8 to 18 pieces of veneer or thin plywood can often be scrolled out simultaneously, depending on the thickness of the

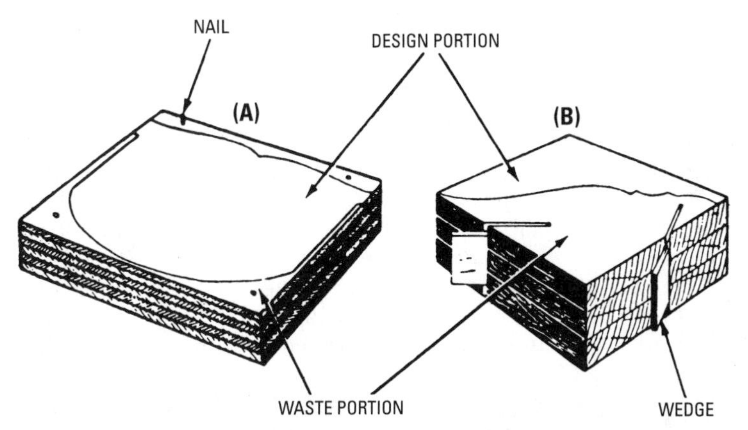

Figure 7-14 The methods of assembly commonly used in multiple-sawing operations: (A) Nails are driven into the waste portions of the design to hold the parts together while being sawed; (B) the stock is wedged together while being sawed.

stock. Some favor the idea of tacking pieces together lightly before sawing, but a more satisfactory system is to cut two or more $^{3}/_{4}$-inch slits in their edges after stacking them up evenly on the saw table. A hardwood wedge driven into each of these slits will serve to hold the stock together while it is being processed (Figure 7-14). Some workmen scroll eight or more pieces of $^{1}/_{2}$-inch plywood at one time without fastening the plies together in any manner. This procedure calls for considerable dexterity, and it is really not advisable to attempt it on work where great accuracy is essential.

Pointers on Band-Saw Operation

The operator should be able to adjust the band saw. This is to obtain the maximum quantity, as well as the best possible quality from the band saw. It is necessary that the operator understand its operation. Prior to the actual sawing, the operator must be carefully versed in the various safety features with which the band saw is equipped. The operator should be familiar with the removal of the wheel guards, and before operating the band saw, should make certain that they are securely fastened. Some band saws are equipped with a braking device whereby the drive wheel may be stopped quickly for blade changes. Some types are provided with an automatic brake that instantly stops the wheels if the blade breaks.

It is a good idea to use the widest blade possible. That is, give due consideration to the minimum radius to be cut on a particular class of work. A rule of thumb used by many is that the width of the blade should be one-eighth the minimum radius to be cut. Therefore, if the piece on hand has a 4-inch radius, the operator would select a $^1/_2$-inch blade. This rule should not be construed to mean that the minimum radius that can be cut is eight times the width of the blade, but rather that such a ratio indicates the practical limit for high-speed band-saw work.

Where the band saw is operated continuously, the blade tension may have to be increased gradually during the day, because the heat of the blade will cause it to expand and stretch. As the blade expands, tension will decrease, and cutting will become more difficult. At the end of the day's work, the blade tension should be relieved, since the blade will contract as it cools and may fracture if the tension is too great. However, this should not be a problem in occasional use.

The operator should also be able to select the correct blade for the class of work at hand. As previously mentioned, the widest blade possible should be used, taking into consideration the radius of the curves to be cut. The reason for this is that on straight or gently curving portions of work, it is much easier to follow the contour with a wide blade, since narrow blades often have a tendency to wander.

Band-saw blades are commonly classified as 4-, 5-, 6-, or 7-tooth blades. This designation refers to the number of teeth per inch of blade length (Figure 7-15). Where smooth cuts are desired, the 6- or 7-tooth variety should be used. Where speed is of more importance than the smoothness of a cut, a 4-tooth blade should be employed, because its larger teeth will cut more rapidly.

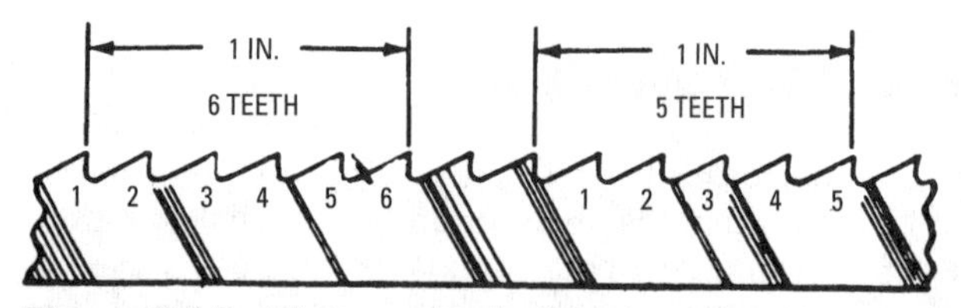

Figure 7-15 The method of designating the number of teeth per inch.

Jigsaw

The *jigsaw* differs in anatomy from the band saw, although the type of work for which it is designed is quite similar. The jigsaw is more adaptable than the band saw for cutting small, sharp curves simply because much smaller and finer blades may be used. Inside cutting is also better accomplished on the jigsaw, because the blade is easily removed and inserted through the entrance hole bored in the stock.

Construction

The jigsaw (Figures 7-16 and 7-17) consists essentially of a base or frame, a driving mechanism, a table, a tension mechanism, guides, and a saw blade. The driving mechanism in a jigsaw has a motor-driven wheel that is connected by a steel rod to a bar (called the *cross-head*) that moves up and down between two vertical slides. This arrangement converts the rotating motion of the motor into a reciprocating (up-and-down) movement of the blade and provides an efficient cutting action.

The table built around the blade is usually designed for tilting at angles up to 45°. The size of the jigsaw is generally expressed in terms of the *throat opening* (that is, the distance from the blade to the edge of the supporting arm). The distinguishing feature of saws of this type, as previously noted, is that the blade moves with a reciprocating motion instead of moving continuously in one direction, as in the case of table and band saws. Accordingly, only one-half the distance traveled by the saw blade is effective in cutting.

With the jigsaw and its numerous attachments, very few cutting operations cannot readily be accomplished. The primary uses of the jigsaw are for cutting out intricate curves, corners, and so on, as in wall shelves, brackets, and novelties of wood, metal, and plastics (Figure 7-18).

Operation

As mentioned, the jigsaw is primarily used for making various types of intricate wood cuttings that cannot readily be made on the band saw. The chucks are generally made to accommodate various sizes of blades, which are inserted with the teeth pointed in a downward direction. In operation, the front edge of the guide block should be in line with the gullets of the teeth and should be fastened in that position. The guide post is then brought down until the hold-down foot rests lightly on the work to be sawed. For most work, the operation of the jigsaw does not differ in any important respect from that of the band saw.

The speed of the jigsaw is generally determined by the material to be cut, as well as the type of blade used, in addition to the skill of the operator. Various speeds of from 650 to 1700 cutting strokes per minute may be selected by the use of the proper step on the cone

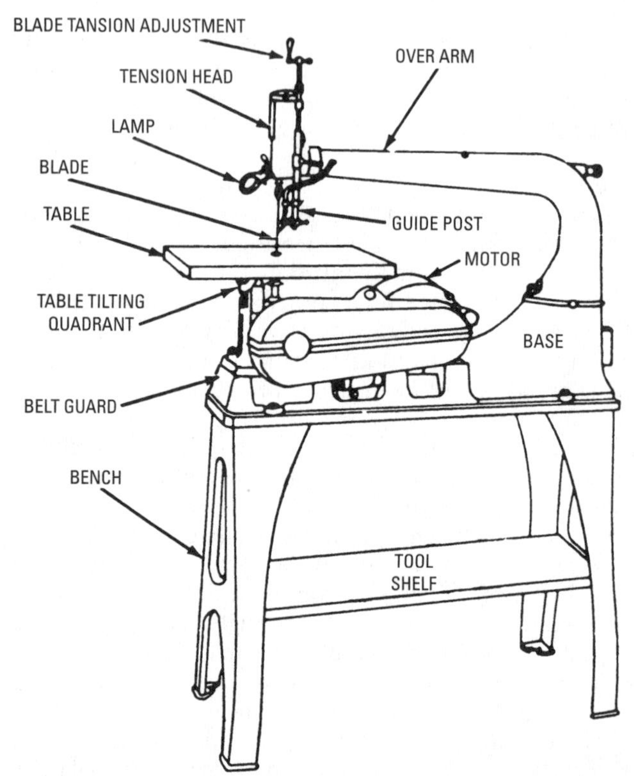

Figure 7-16 The component parts of a typical jigsaw. A jigsaw of this type uses a 6-inch blade, but it can accommodate smaller and larger blades. The table can be tilted 45° either way and is equipped with a graduated quadrant showing the exact number of degrees of tilt. The upper head adjusts on dovetail ways by means of a hand crank and a knurled locknut. Blade tension is shown on a scale and may be regulated for minimum vibration while the machine is running. The drive mechanism is of the reciprocating type with link and counterbalanced crank. The air pump is powered by the main drive shaft and keeps the cutting line clear at all saw-blade speeds. The guides are adjustable for front and side sawing.

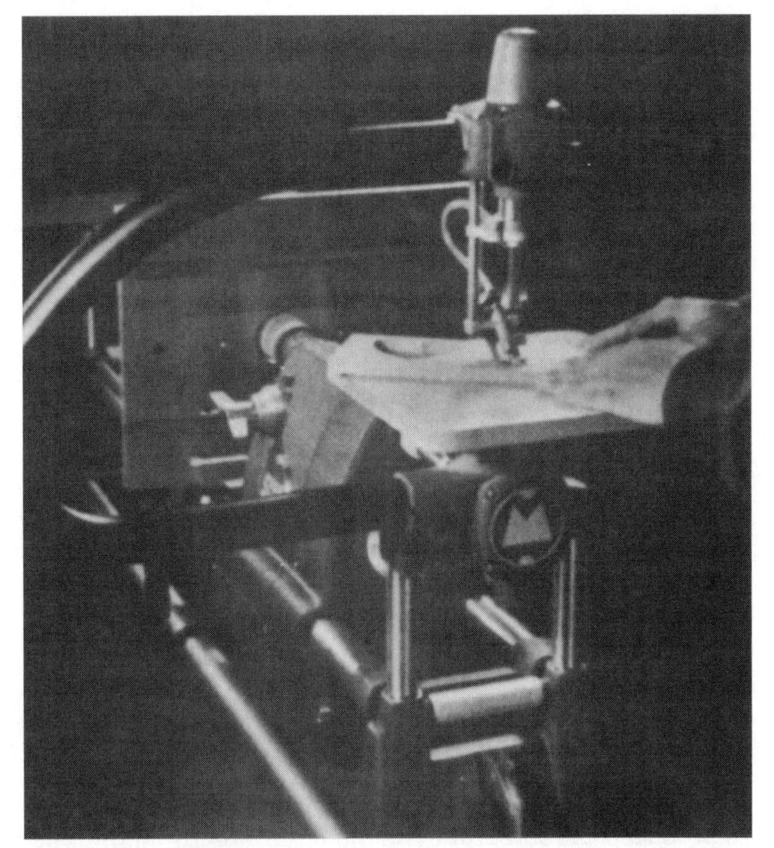

Figure 7-17 This jigsaw is part of a combination tool. A wide throat allows manipulation of large stock. *(Courtesy of Shopsmith)*

pulley. Jigsaw blades vary a great deal in length, thickness, width, and fineness of the teeth. All blades, however, may be grouped under two general classifications:

- Blades gripped in both the upper and lower chucks
- Blades held in the lower chuck only

The latter types of blades are known as *saber blades* (Figure 7-19), whereas the former are called *jeweler's blades*. The jeweler's blades are useful for all fine work where short curves predominate, while saber blades are faster cutting tools for heavier materials and

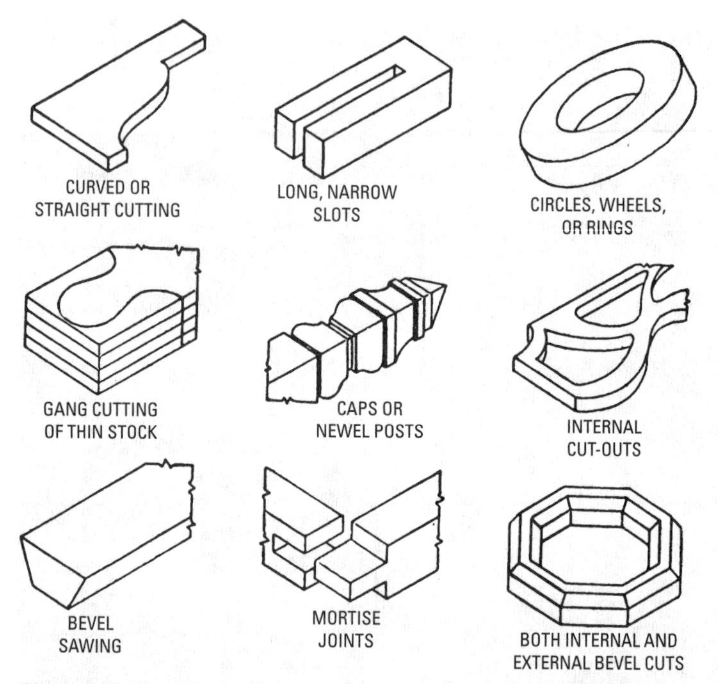

Figure 7-18 Various items produced by means of jigsaw cutting. One of the most important steps in cutting any shape from wood is that of marking pattern shapes on the wood to be cut. This is usually done by drawing the pattern with the aid of suitable squares or by the use of an outline projector. The pattern so produced can sometimes be mounted directly on the wood as a cutting guide, or else the pattern may be produced directly on the work by means of carbon paper.

medium curves. When it is desirable to make inside cuts, a starting hole is normally drilled at a suitable location.

Wood Lathe

A *wood lathe* is a stationary power tool used for shaping wood. The wood revolves in the lathe while a sharp-edged cutting tool held in the hand and supported by a slide rest is pressed against it. Since the wood is revolving while being cut, the operation is termed *woodturning*.

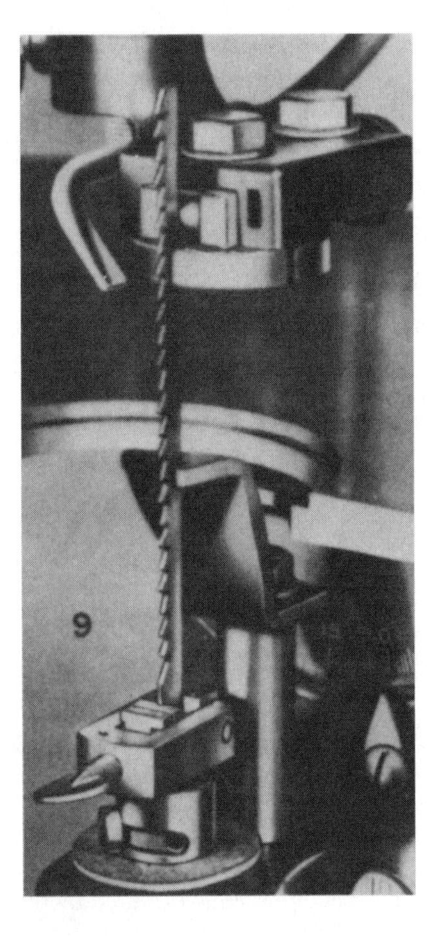

Figure 7-19 Saber saw blades might be used without being chucked in at the top. This can be convenient when making certain cuts. *(Courtesy of Rockwell)*

Construction

The principal parts of a wood lathe are the *bed*, *headstock*, *tailstock*, and *tool rest* (Figure 7-20).

The bed usually consists of a heavy casting with two parallel ways and is supported by cast-iron legs on a table to bring the work up to the desired height. On some lathes, the bed and ways are combined in a single cylindrical pipe (Figures 7-20 and 7-21).

- The ways carry a rigid headstock at the left end and a tailstock at the right end.

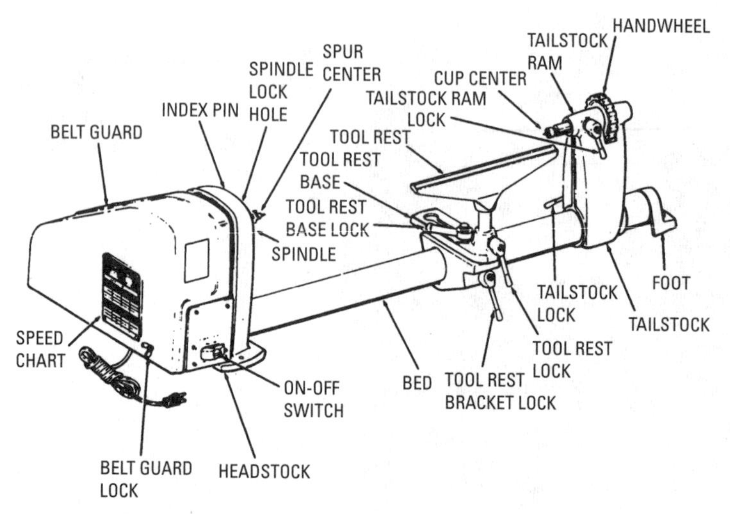

Figure 7-20 Essential parts of the wood lathe.

- The tailstock can slide on the ways and can be secured at any point by tightening the tailstock clamp.
- The headstock is bolted to the bed. It contains the driving mechanism of the lathe. The driving mechanism is a hollow spindle supported between two bearings.
- The spindle is revolved either by a step-cone pulley or by an individual motor, in which case the spindle is directly connected to the motor.

A *spur center* (Figure 7-22A) is fitted into the spindle and engages one end of the wood to be turned. The other end is secured by the tail center (Figure 7-22B or 7-22C). As the spindle turns, power is transferred through the spur center turning the wood.

Both the tool post and the tailstock can be clamped to the bed at any point, allowing for variation in the length of wood to be turned. By resting a sharp-edged cutting tool against the T-shaped tool rest, the wood is shaved off and the surface is reduced to a circular form.

Lathe Speeds

The lathe speed may be regulated in several ways, depending on the method of drive. When the lathe is driven from a set of step pulleys, the speed at the cutting edge will range from 100 to 2500 feet per minute, depending on the type of turning to be done.

Figure 7-21 You should mount the lathe on a sturdy table, such as the one shown. *(Courtesy of Sears, Roebuck & Co)*

When the headstock is directly motor-driven, the speed is usually regulated by a rheostat. Or it can be regulated by a special switching arrangement that provides for up to four speeds, usually from 500 to 3600 rpm. No definite rule for lathe speeds can be laid down, however, because of the large variations in diameter that often occur in the piece to be turned. Too slow a speed not only is a great waste of time, but will also leave a rough finish on the wood. Too fast a speed may damage the cutting tools. Always remember that the largest diameter of the material will determine the lathe spindle speed.

Starting and Stopping the Lathe

Prior to starting the lathe, the adjustments and clamps should be tested to ensure their workability, and the work should be revolved by pulling the belt by hand to make sure that the work clears the

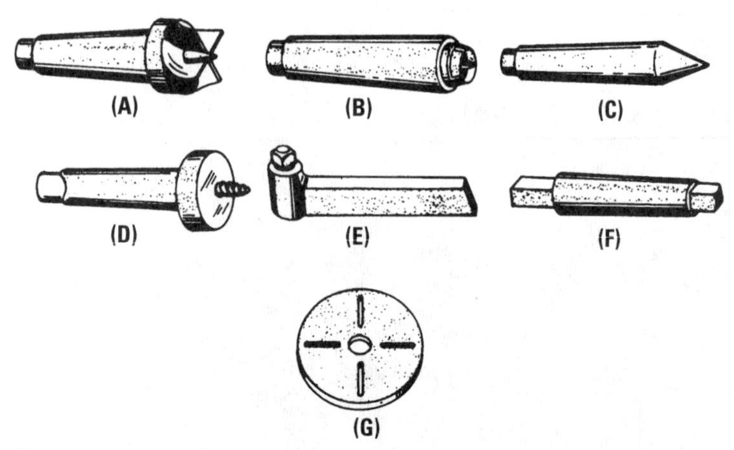

Figure 7-22 Lathe attachments: (A) spur drive center; (B) cup center; (C) cone center; (D) screw center; (E) tool holder; (F) adapter; (G) faceplate.

tool post. It is better to start turning at a safe speed and increase it, if necessary, after the work is rough-turned and is running true.

The lathe should never be run at high speed if any part of the work is off-center or out of balance. If the work to be turned is out of balance, it should be counterbalanced by suitable weights, although this may not always be possible. To stop the machine, the belt is shifted, or the motor is disconnected and braking is accomplished by placing a hand on the pulley. Large faceplate work, however, should not be stopped too suddenly in this way, because there is danger of unscrewing the plate from the spindle.

A smooth piece of work can be safely stopped by braking on the work itself, using a handful of shavings or smooth turnings to prevent burning. However, the lathe should never be stopped by placing a hand on the back or edge of the faceplate, since there are likely to be projecting objects on it that may cause injury to the operator.

Lathe Accessories

Lathe attachments (Figure 7-22) consist of several tools or accessories necessary to perform the work properly. They are *lathe centers, drivers, screw chucks, center chucks,* and *faceplates*. Accessories include various *turning* and *measurement* tools.

Lathe Centers

The function of the lathe centers is to hold the work and revolve it between the spindles. They are of three principal types—the *spur* or *live center*, the *headstock* or *tailstock center*, and the *cup center*.

The spur or drive center is used in the live spindle for driving small and medium-sized pieces that are to be turned between the centers. The spur center is tapered to fit the hole in the spindle, and the driving end has a point and either two or four spurs to engage the wood. The spur center is inserted into the piece to be turned by setting the point in the center and driving the spurs in by striking the end of the center with a mallet. The center is removed from the lathe by pushing a rod through the hole in the spindle.

The tailstock center is used to support the right-hand end of the work and is generally self-discharging. It may be removed from the spindle by drawing the spindle back with the hand wheel as far as it will go. This action will automatically push the center out of the spindle.

The cup center is also a tailstock center and has a thin, circular steel edge around a central point. It is more accurate than the cone center and does not split the wood so easily. Since the dead or tailstock center does not revolve, the end of the stock that turns on it should be well-oiled to prevent burning by friction.

Lathe Drivers

Drivers are devices (other than the spur center) used for revolving the work. There are several forms available, with some made for use with a small slotted faceplate and dog (as in machine-shop turning), the center plate having a projection to which the dog is fastened. Another form of driver is one in which the center has a square shank over which the dog fits.

Screw centers are small faceplates with a single screw in the center and are used for turning small pieces. Some screw centers hold the central screw firmly and have an attachment for regulating the projection of the screw through the face.

Faceplates are holding devices used for work that cannot readily be driven by any of the foregoing methods. They contain screw holes countersunk on the back for the screws used in fastening the wooden face or chuck plates. The smaller sizes are frequently provided with a recess in the center for rechucking purposes. They are made in various sizes, depending on the size of the lathe and the stock they must support. Objects such as circular disks, bowls, and trays are always turned on faceplates.

Woodturning Tools

Woodturning tools (Figure 7-23) are used for cutting and scraping purposes, the most common of which are various types of gouges, chisels, and parting tools. *Gouges* are beveled on the outside or convex side, and the length of the bevel is about twice the thickness of the steel. *Skew chisels* are beveled on both sides, and their cutting edges form a 60° angle with one side of the chisel. *Parting tools* or *cut-off tools* are thicker in the center of the blade than at the edges and, thus, will not bind or overheat when a cut is made. This tool is used for making narrow cuts to a given depth or diameter.

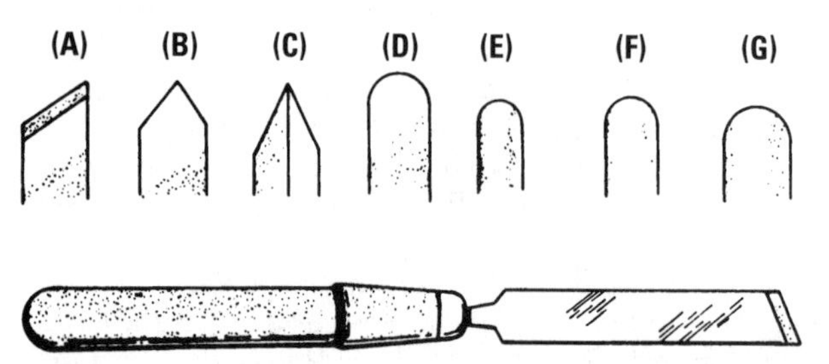

Figure 7-23 Woodturning tools: (A) skew chisel; (B) spear-point chisel; (C) parting tool; (D) round-nose chisel; (E) gouge; (F) gouge; (G) gouge.

Measurements

The ability to take accurate measurements plays an important part in woodturning work. This ability can be acquired only by practice and experience. All measurements should be made with an accurately graduated scale.

An experienced operator can take measurements with a steel scale and calipers to a surprising degree of accuracy (Figure 7-24). This is accomplished by developing a sensitive caliper feel and by carefully setting the calipers so that they split the line graduated on the scale.

Outside Calipers

A good method for setting an *outside caliper* to a steel scale is shown in Figure 7-25. The scale is held in the left hand and the caliper in the right hand. One leg of the caliper is held against the end of the scale and is supported by the finger of the left hand while the adjustment is made with the thumb and first finger of the right hand.

The proper application of the outside caliper when measuring the diameter of a cylinder or a shaft is shown in Figure 7-26. The caliper

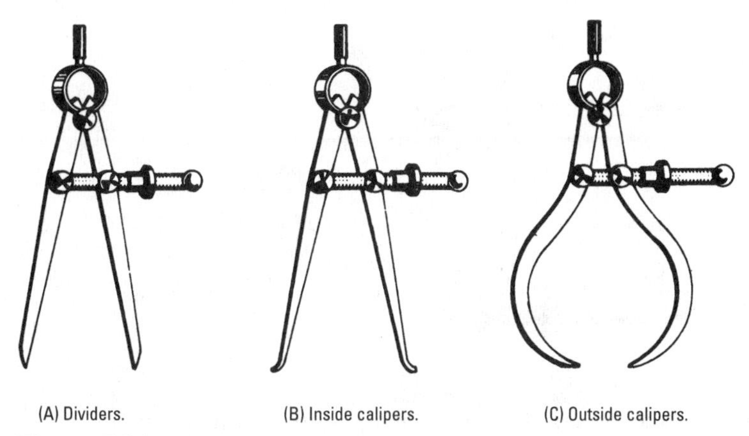

(A) Dividers. (B) Inside calipers. (C) Outside calipers.

Figure 7-24 Measurement tools.

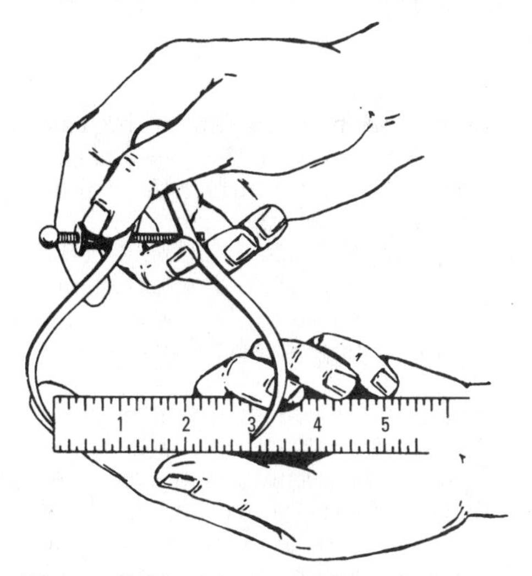

Figure 7-25 Method of setting an outside caliper to a steel square.

is held exactly at right angles to the centerline of the work, and is pushed gently back and forth across the diameter of the cylinder to be measured. When the caliper is adjusted properly, it should slip easily over the shaft of its own weight. Never force a caliper, because it may spring and the measurement may not be accurate.

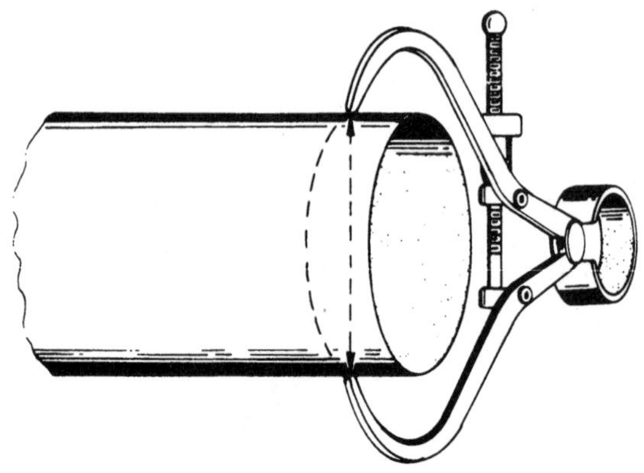

Figure 7-26 Method of measuring with an outside caliper.

Never grip the caliper too tightly. The sense of touch will be very much impaired.

Inside Calipers

To set an *inside caliper* for a dimension, place the end of the scale against a flat surface and the end of the caliper at the edge and end of the scale. Hold the scale square with the flat surface. Adjust the other end of the caliper to the required dimension (Figure 7-27).

To measure an inside diameter, place the caliper in the hole, in the position shown by the dotted line in Figure 7-28, and raise the hand slowly. Adjust the caliper until it will slip into the hole with a very slight drag. Be sure to hold the caliper square across the diameter of the hole.

In transferring a measurement from an outside caliper to an inside caliper, the point of one leg of the inside caliper rests on a similar point of the outside caliper (Figure 7-29). Using this contact point as a pivot, move the inside caliper along the dotted line shown in Figure 7-29, and adjust with the thumbscrew until you feel that the measurement is just right. The *hermaphrodite caliper* (Figure 7-30) is set from the end of the scale exactly the same as the outside caliper.

The accuracy of all contact measurements is dependent upon the sense of touch or sense of feel. The caliper should be delicately and lightly held in the finger tips—not gripped tightly. If the caliper is gripped tightly, the sense of touch is lost to a great extent.

Centering and Mounting Stock

Wood stock to be turned must be properly marked. That is, the true centers must be obtained prior to turning. In square stock, the

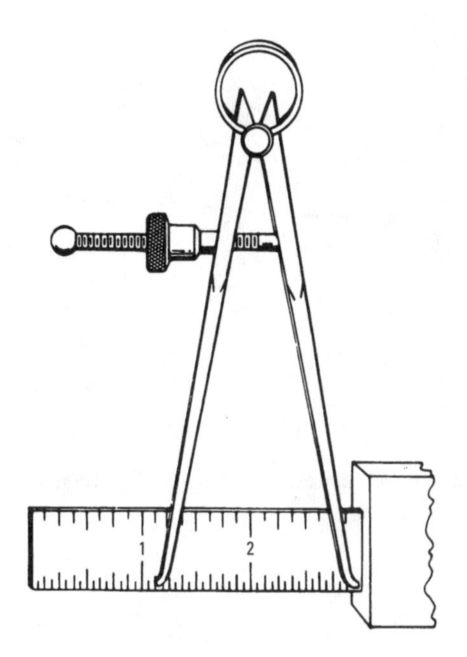

Figure 7-27 Method of setting an inside caliper.

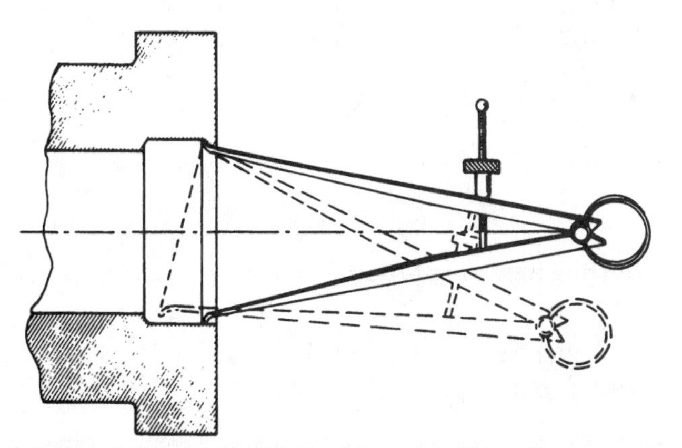

Figure 7-28 Measuring with an inside caliper.

center is usually determined by the diagonal method, which consists of holding the center head of a combination square firmly against the work and drawing two lines at right angles to each other and close to the blade across each end of the work.

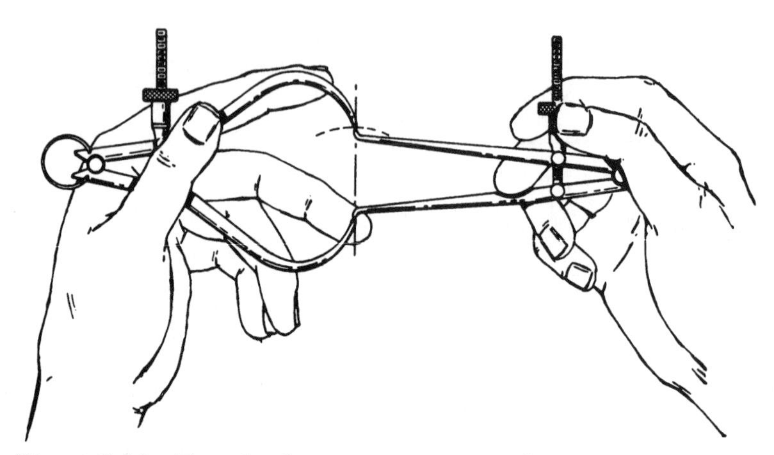

Figure 7-29 Transferring a measurement from an inside to an outside caliper.

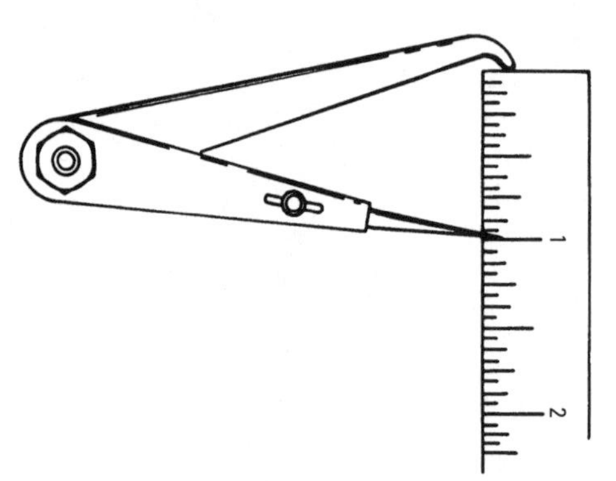

Figure 7-30 Setting a hermaphrodite caliper.

After the work center has been marked, the work can be mounted in the lathe (Figure 7-31):

1. The stock is first pressed against the spur or live center so that the spurs enter the grooves previously marked.
2. Next move the tailstock up to about 1 inch from the end of the stock and lock it in this position.
3. Then advance the tailstock center by turning the feed handle until the center makes contact with the work.

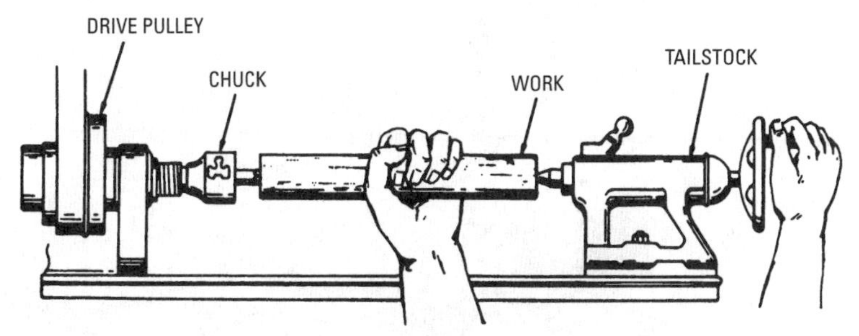

Figure 7-31 Method of mounting work in a lathe.

4. Continue to advance the center while slowly rotating the stock by hand.
5. After it becomes difficult to turn the stock, slack off on the feed about one-quarter turn and lock the quill spindle. The stock is now ready for turning.
6. The tool rest is next mounted in position so that it is level with the centers and about $^1/_8$ inch away from the stock.
7. Clamp the tool rest in position and revolve the stock by hand to make certain it has sufficient clearance.

Typical Woodturning Operations

The most common woodturning operations include such simple steps as roughing off, paring or finishing, squaring off ends, and making concave or ring cuts. In a roughing-off cut, place a large gouge on the rest so that the level is above the wood and the cutting edge is tangent to the "circle of cut." In adjusting the rest for this, the handle of the gouge should be well down. Roll the gouge over slightly to the right so that it will shear instead of scrape the wood. Lift the handle slowly forcing the cutting edge into the wood. Remove the corners of the wood first, and then the intervening portions. The tool should be held at a slight angle to the axis of the wood being turned, with the cutting end in advance of the handle (Figure 7-32).

Follow these steps:

1. In making a paring or finishing cut, lay a skew chisel on the tool rest with the cutting edge above the cylinder and at an angle of about 60° to the surface.
2. Slowly draw the chisel back and, at the same time, raise the handle until the chisel cuts about $^1/_4$ to $^3/_8$ inch from the heel.

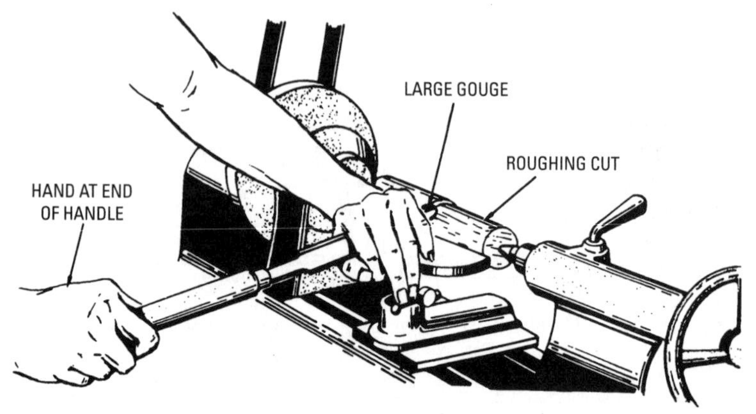

Figure 7-32 The correct way to hold a gouge in making a rough cut on the lathe.

3. Begin the first cut about 1 to 2 inches from either end, pushing toward the other end.
4. Then begin at the first starting point and cut toward the other end, thus taking off the rings left by the gouge, and cut down to where the scraped sections are just visible.
5. Take a last cut to remove all traces of the rings and to bring the wood down to a uniform diameter from end to end.
6. To test for smoothness, place the palm of the hand, with the fingers extended, lightly on the back side of the wood away from the tool rest.
7. After the work has been reduced to a cylinder, the ends may be squared to mark the end of the turning. This is commonly accomplished by placing the 1/4- or 1/2-inch skew chisel on the tool rest and bringing the cutting edge next to the stock perpendicular to the axis, as shown in Figure 7-33.
8. The heel of the chisel is then slightly tipped from the cylinder to give clearance.
9. Raise the handle and push the chisel toe into the stock about 1/8 inch outside the line indicating the end of the cylinder.
10. Swing the handle still further from the cylinder and cut a half "V" to give clearance and prevent burning.
11. Continue this process until the cylinder is cut through to about a 1/16-inch diameter.

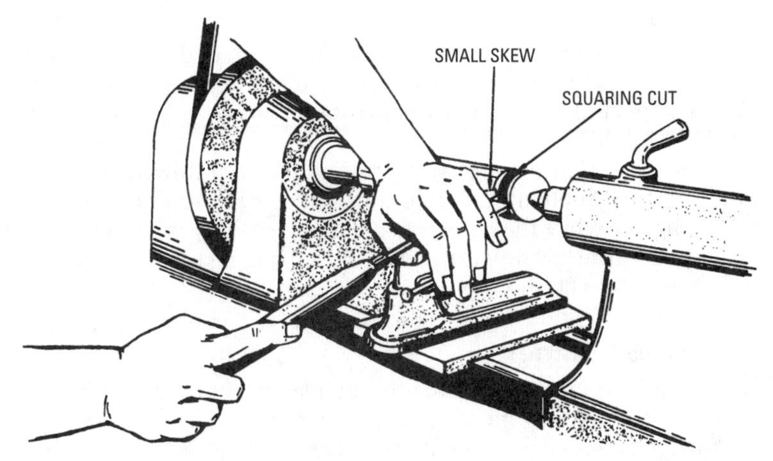

Figure 7-33 A skew chisel is held in this manner to square the ends of the work.

12. When making concave cuts, place the gouge on the tool rest with the cutting edge well above the wood.
13. Roll the tool to the side until the bevel at the cutting point is perpendicular to the axis of the cylinder.
14. Slowly raise the handle to force the gouge into the wood. When the gouge bites into the wood, force it forward and upward by slightly lowering the handle while rolling it back toward its first position.
15. Reverse the position of the gouge and make a similar cut from the other side to form the other half of the semicircle.
16. Ring cuts are made by placing a 1/4- or 1/2-inch small skew chisel on the tool rest with the cutting edge above the cylinder and the bevel tangent to it.
17. Draw the chisel back and raise the handle, bringing the heel of the chisel into contact with the cylinder at the marked line.
18. The chisel is then moved to the right (if cutting the right side of the bead) while being continually tipped to keep the lower bevel tangent to the revolving cylinder and to the head at the point of contact.
19. Continue the cut until the bottom of the convex surface is reached.

Polishing

The polishing of turned work in the lathe is a great deal easier than applying polish to a flat surface. Here, the polishing operation is done with a cloth while the work is rotating in the lathe and after it has been sanded.

In sanding, first use a coarse grade of paper, followed by a fine grade. Before applying the polishing cloth, the wood may be varnished lightly while the lathe is not running, taking care to wipe off all surplus varnish. The varnish will assist in giving the surface a fine polish when the cloth is applied.

Thickness Planer

In general, a *planer* is a machine for planing or *surfacing* wood (especially boards) by means of a rapidly revolving cutter that chips off the rough surface in minute shavings as the piece to be planed is passed over the revolving cutter by power feed, leaving a smooth or finished surface.

Lumber thus treated or planed is said to be *dressed*. Lumber may be dressed on one side, which is abbreviated *S1S* meaning *surfaced one side*. When dressed on two sides, the designation is *S2S*. *S4S* is used when both sides and edges are dressed.

The purpose of surfacing the wood is to smooth the uneven and splintery surface of rough-cut lumber. The thickness planer also makes both sides of a board parallel with each other so the board has an even thickness throughout.

Wider boards are passed through the machine one at a time. Narrower boards may be fed through side-by-side, staggering them lengthwise to make handling easier (Figs. 7-34 through 7-37).

Snipe is a slight reduction in thickness at the end of a planed board. It occurs when the board drops off bed rollers. This means it is no longer held by the in-feed roller. Prevent snipe by holding up the free end of the board or by feeding a short board immediately before and after the board being planed.

Jointer

The jointer is similar to the planer except that it has only one cutter head located below the table, and the feed is usually by hand. Its primary use is to cut a true face and edge on stock that is warped, twisted, or has other irregularities. The table is in two sections, the *in-feed section* being raised or lowered to control the thickness of the cut. The *out-feed part* of the table is raised or lowered by a unit to control the thickness of the finished piece. By tilting the fence,

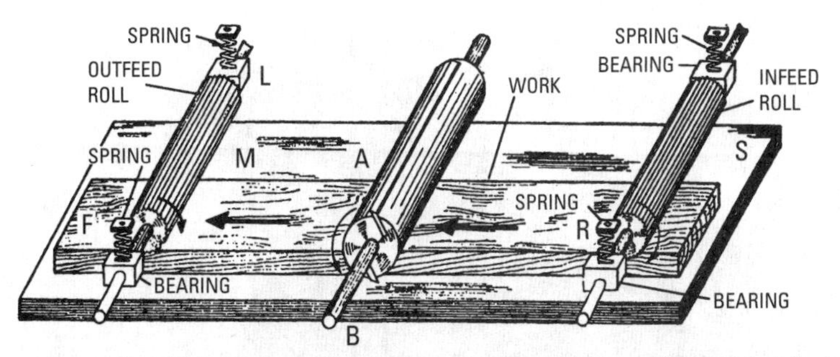

Figure 7-34 Principles of planer operation. The cylindrical cutter head (A), containing the cutter knives, rotates at high speed. The head's shaft is geared to the in-feed roll (S) and out-feed roll (M) shafts, which rotate at a slow speed. These rolls are pressed down on the board by the springs (F,L,R,A) working through the bearings as shown. The corrugated rolls grip the board and slowly push it along the table in the direction of the arrow. First the board is fed into the cutters by roll (M). The rotating knives cut into the board leaving a smooth surface. Then roll (S) continues feeding after the end of the board leaves roll (S) and the cylinder.

bevel edges can be cut. This machine can also be used for tapering, end planing, and rabbeting.

Construction

The essential parts of a jointer (Figure 7-38) consist of a *base* or frame supporting the assembly, *cutter head, tables* (or *bed*), *guide fence*, and *table-adjusting hand wheels.*

The cutter head houses two or more high-speed steel knives and revolves between the front or in-feed table and the rear or out-feed table. The tables may be raised or lowered by means of hand wheels to regulate the depth of cut.

A guide fence extends along both tables to hold the work as it is advanced toward the cutting knives. The fence may be tilted 45° either way and may be moved crosswise along the jointer table. The cutter head is generally provided with a guard cover as a safety measure.

Adjustments

One of the most important adjustments on the jointer is the relation of the rear table to the cutter head. To do satisfactory work, this

Figure 7-35 Planing a wide board. The protective housing has been removed to show the cutter head and feed rolls. The rolls on this planer are smooth and less likely to mar the new surface than corrugated rolls like those shown in Figure 7-34. *(Courtesy of Belsaw)*

table must be exactly level with the knives at their highest point of rotation. To make this adjustment, follow the manufacturer's instructions. Figure 7-39 shows the correct and incorrect positions of the rear table.

The in-feed table, on the other hand, must be somewhat lower than the rotating knives, the exact height of which depends upon the depth of the cut. The difference in height between the two tables is equal to the depth of the cut. Improperly adjusted tables will result in the work being cut on a taper.

When knives become dull or nicked, it is necessary either to replace them or to sharpen them. Jointing knives may be sharpened and brought to a true cutting circle by jointing their edges while the cutter head is revolving. In performing this operation, a fine cutting oilstone is placed on the out-feed table, and the table is lowered until

Figure 7-36 Molding on the planer. Short molding cutters with a curved profile are fitted into the cutter head. A curved profile is cut onto the edge of the board as it passes through the planer.

(Courtesy of Belsaw)

the knives just touch the stone at their highest point of rotation. It is easier for the neophyte to replace blades.

Operation

Although a great number of woodworking operations may be performed on the jointer, the most common are edge planing, planing of ends, and surfacing (Fig. 7-40 through Figure7-42). Among these, edge-planing or jointing an edge is the simplest and most common operation. Assuming the machine to be properly adjusted and the guide fence square with the table, all that is necessary is to feed the work over the rotating cutter heads. The left hand presses the work down on the rear table so that the cut surface will make good contact with the table. The right hand exerts no downward pressure on the front table, but simply advances the work over the knives. Both hands, however, exert pressure on the work to force it against the guide fence to obtain the proper edge.

Figure 7-37 Ripping on the planer. A ripping attachment with a circular saw blade is bolted to the front of the planer. It rips the board in two before it is planed. Combined functions like this save time on long runs in the production shop. *(Courtesy of Belsaw)*

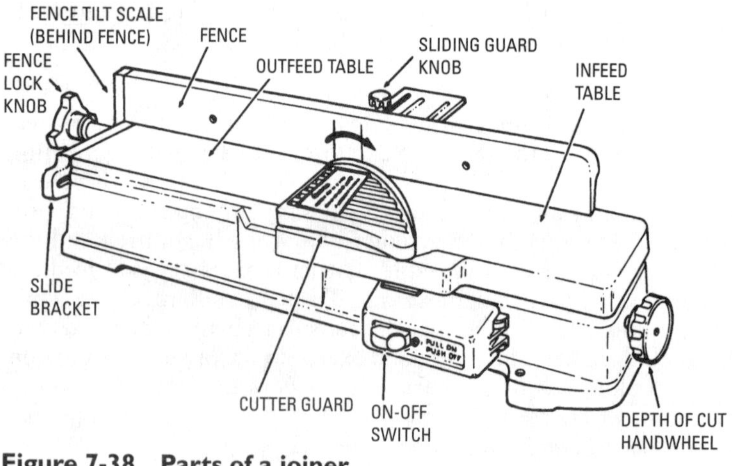

Figure 7-38 Parts of a joiner.

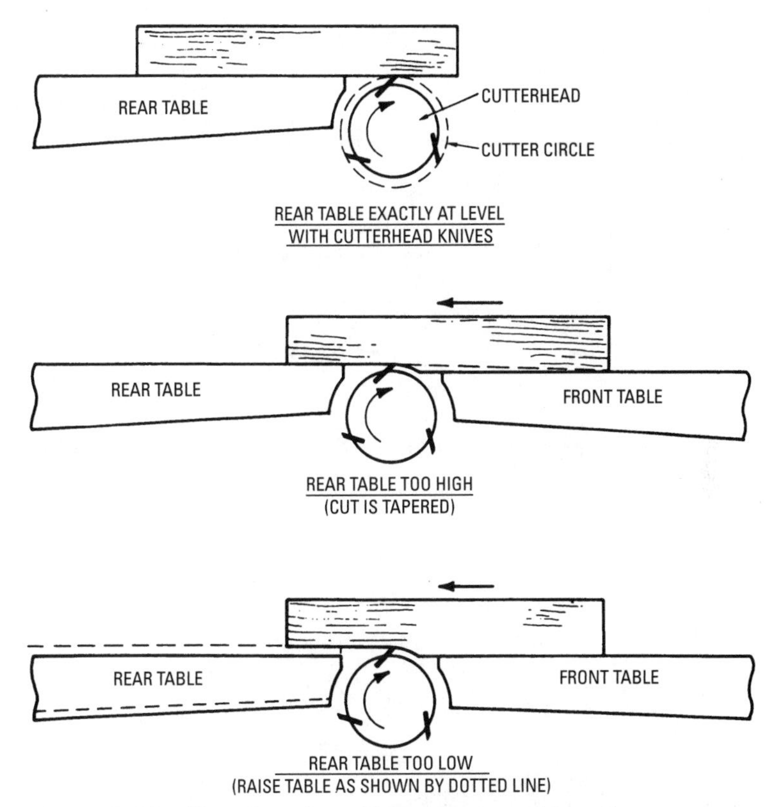

Figure 7-39 Showing the effect of rear-table adjustment on the jointer. The rear table must be adjusted until it is exactly level with the cutter knives at their highest point of rotation. The front table is lowered by the amount of the desired cut.

Edges on thin work (such as veneer) may also be planed on the jointer provided the work is held firmly between two suitable boards or planks. In most edge-planing operations of this type, special clamps are used to hold the work in place.

Planing across the ends of boards may be performed in a manner similar to planing with the grain. The necessary precaution consists in taking only light cuts each time because of the tendency of the grain at the end of the board to split if thick cuts are attempted. To avoid splitting the end wood, some operators make only a short cut at one edge, then reverse the wood and complete the planing from the opposite edge.

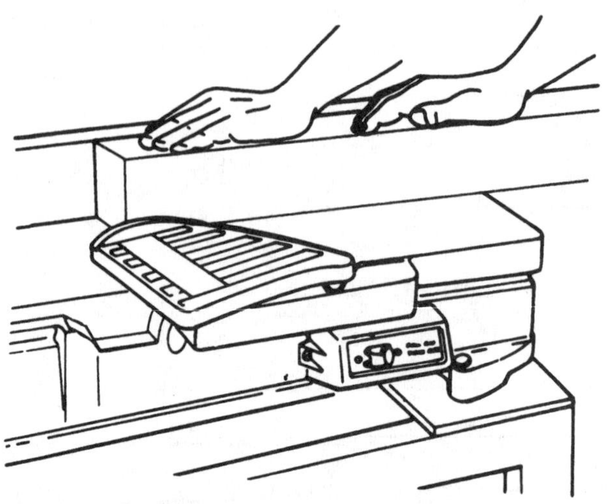

Figure 7-40 Hand-over-hand motion is best when jointing stock over 3 inches thick.

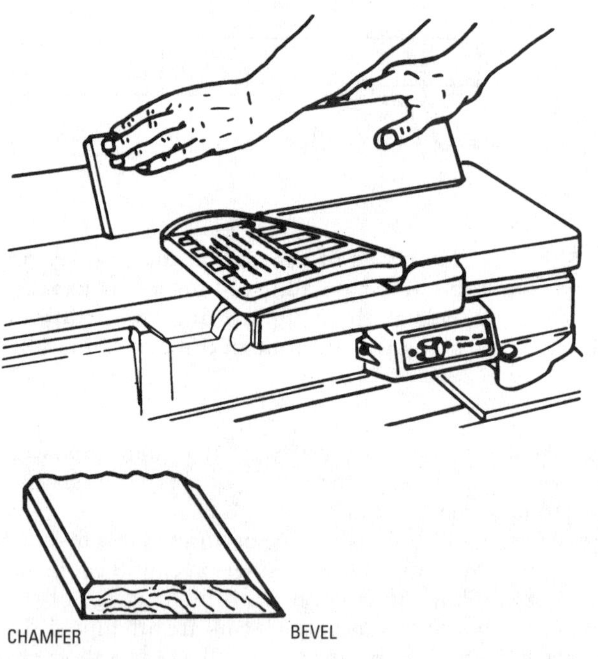

Figure 7-41 Keep fingers together when edge-jointing. Note edged boards.

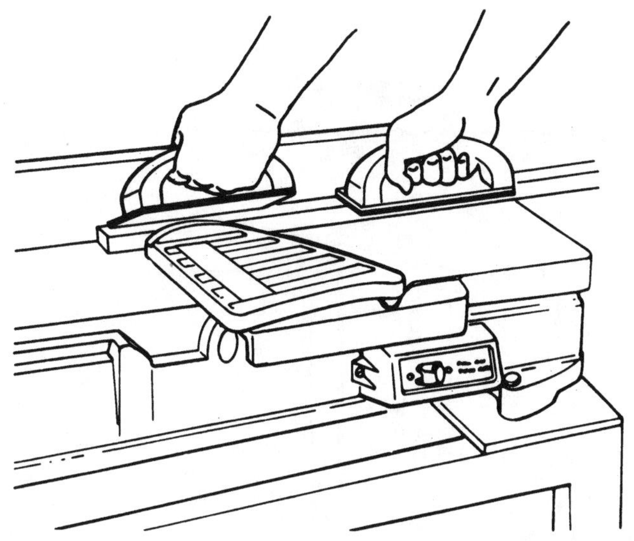

Figure 7-42 When stock is small, stock pusher blocks should be used.

Surfacing the faces of a board is one of the more difficult jointer operations. When planing the faces, the grain pattern in the edges should be noted. To plane the board with the grain, the two sides of the board must be planed in opposite directions.

Shaper

A *shaper* (Figure 7-43) is used mainly to form a curved profile on the edge of stock, and consists essentially of one or more vertical spindles with cutting heads, table, fence, and a base. Generally, a knife with its cutting edge ground to the desired shape of cut is used, but solid cutters milled to the desired shape are also available.

Stock may be cut roughly to shape with a bandsaw or jigsaw before being finished on the shaper. When a number of pieces of the same pattern are to be shaped, special templates or forms are often used to regulate the depth of cut by bearing against the shaper collar.

Construction

Shapers are usually constructed with one or more vertical spindles that carry the cutter heads. Two spindles revolving in opposite directions are perhaps the most general and useful form, since one

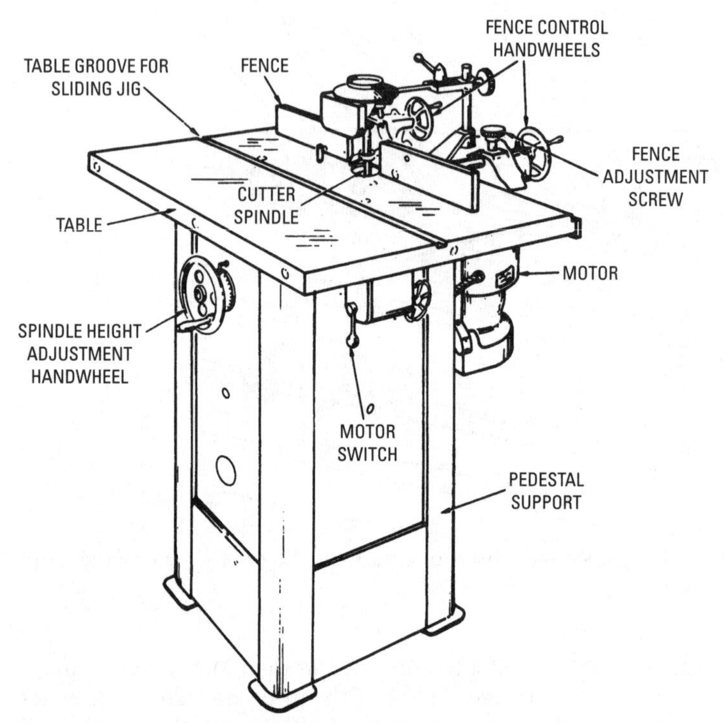

Figure 7-43 A typical shaper.

cutter can cut with the grain of wood running in one direction and the other cutter can operate on wood with the grain running in the other direction without stopping to reverse the machine (Figure 7-44 and Figure 7-45).

The spindle (Figure 7-46) projects above the surface of an accurately planed table. Cutters are attached to this spindle, and the assembly rotates at speeds of 5000 to 10,000 rpm or higher. Because of this high speed, pulleys of about 3:1 ratio are generally necessary when the machine is driven by a standard 3600 rpm motor.

A wide variety of *knives*, *saws*, *collars*, and so on, are used in shaper operations. Shaper knives are of two kinds: the open-face (or flat) knives and the three-lip cutters. The *flat knives* are used mainly on large machines, and have beveled edges that fit into corresponding slots milled in two collars. Knives of exactly the same width are always used, and they must be clamped securely to prevent them from loosening while the machine is in operation.

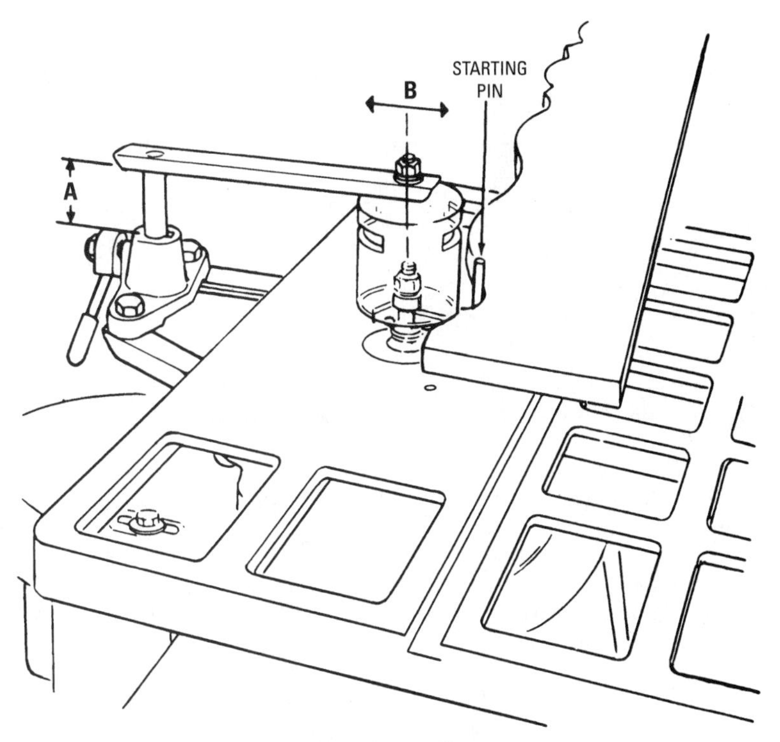

Figure 7-44 One operation of the shaper.

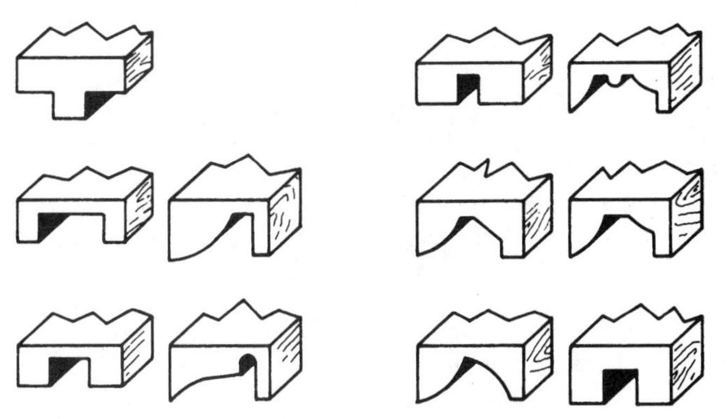

Figure 7-45 Some of the many cuts that can be made on a shaper.

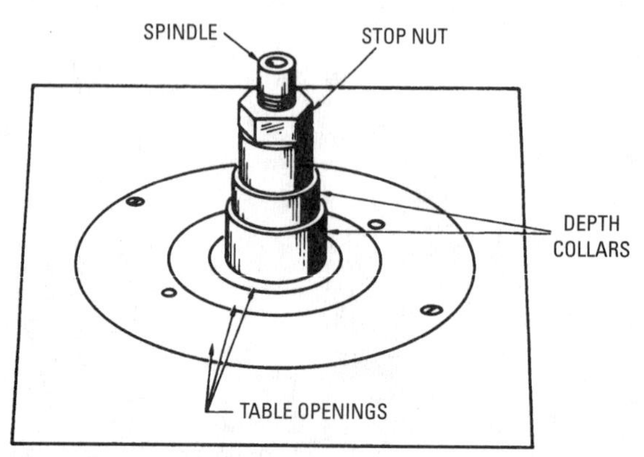

Figure 7-46 The shaper spindle projects through the table.

The *three-lip shaper cutters* have three cutting edges and slip over the spindle shaft. This type of cutter is much safer to work with (Figure 7-47).

Operation

Following are four main methods used to hold or guide the stock against the shaper knives:

- With guides or fences
- Against collars
- With an outline pattern
- With forms

Each one of these methods is widely employed and each is suitable for a certain kind of work (Figure 7-48). The work is held against a fence as it is advanced into the cutter (Figure 7-48A). The position of the fence thus determines the depth of cut. This method is used only for straight cuts. Figure 7-49 shows the adjustment of a shaper fence. The depth of cut is regulated by a collar—a deeper cut requires a collar with a smaller diameter (Figure 7-48B). The edges of an irregularly shaped piece of material can be cut using this method. Figure 7-48C shows the use of an outline pattern. With this method, the pattern rides against the collar to determine the depth of cut. Figure 7-48D shows a form being used to hold the work in position so that it can be advanced to the shaper cutter.

Where a job calls for working the stock directly against an ordinary shaper collar (Figure 7-48), the friction created between the collar and the work has a tendency to burn and mar the material.

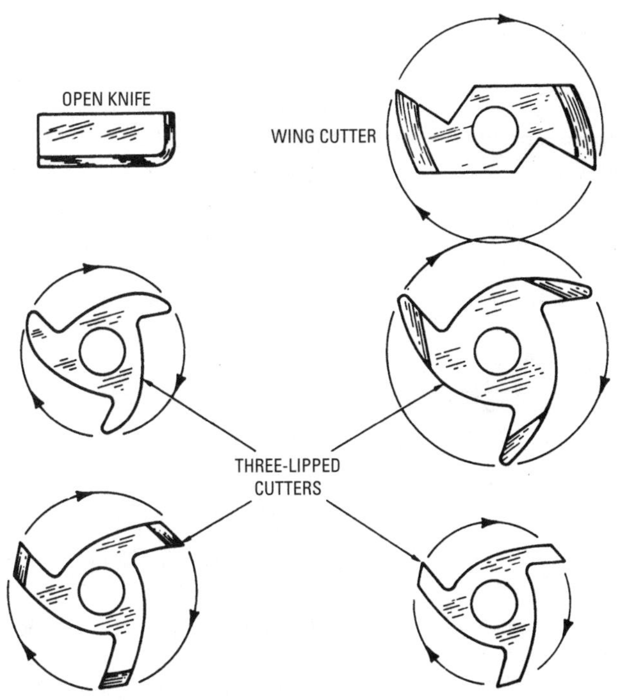

Figure 7-47 Various types of cutters are available for use on a shaper.

This burning can be avoided, however, by using a ball-bearing guide collar in place of the regular type. The outer shell of the ball-bearing collar revolves independently of the inner shell, and thus rolls along with the feed, practically eliminating all wear and friction on any forms or guide patterns being used or on the material being shaped.

Heavy cutting is best accomplished with open-faced knives set in open collars, with the assembly being well-balanced to avoid vibration and render the setup as safe as possible. Whenever it is necessary to work directly against the shaper collar, it is a good idea to rest the stock against a steel guide-pin while easing it slowly to the knives at the beginning of each cut. If the character of the work is such that it is impossible to employ a hold-down device, two or more light cuts should be made where considerable material is to be removed. It is advisable to relieve the cut as much as possible by chamfering off the corners of the stock with a band saw on which the table can be tilted, or on a tilting-blade bench saw. The equipment that is really

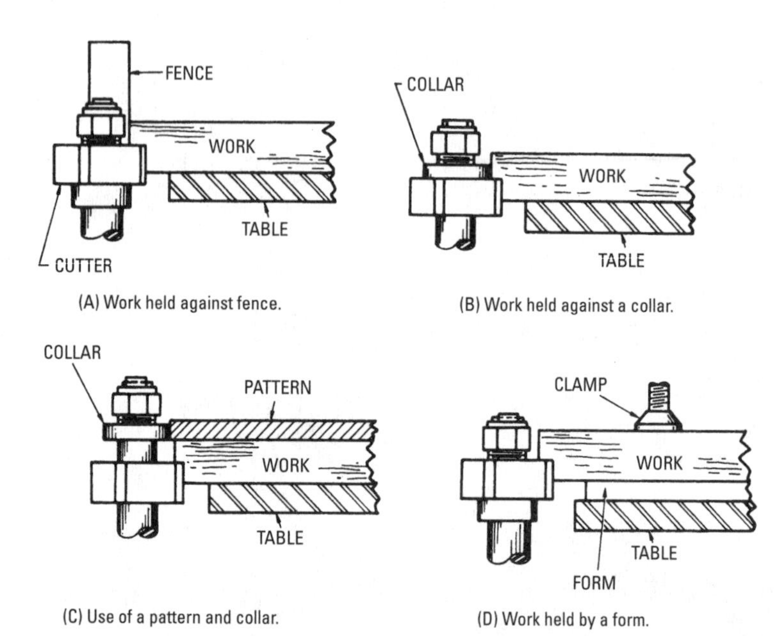

Figure 7-48 Four principal methods used to hold the work against the shaper knives.

best adapted to extra-heavy cutting is the type of shaper with overhead bearings, although some operators prefer a machine having an auxiliary drop table.

In the type of operations employed in the average millwork shop, shaping with guides is a very popular method. The most common types of guides are various fences to suit the particular operation desired. Capable operators keep an ample supply of fences on hand, each usually being adapted to one particular class of work. These generally include one for chamfering, another for grooving, and still others for panel raising, rabbeting, and other setups called for frequently. When using a saw on the shaper spindle for grooving, it is customary to tack a strip of plywood panel stock on the face of the fence, which is then positioned so that the blade cuts through the plywood the proper distance to make the required depth of groove.

When milling short and light work, you can sometimes use the strips of hardwood that serve as guard and hold-down as a fence. This is because it is readily set at any required distance from the cutters by means of the adjusting screws.

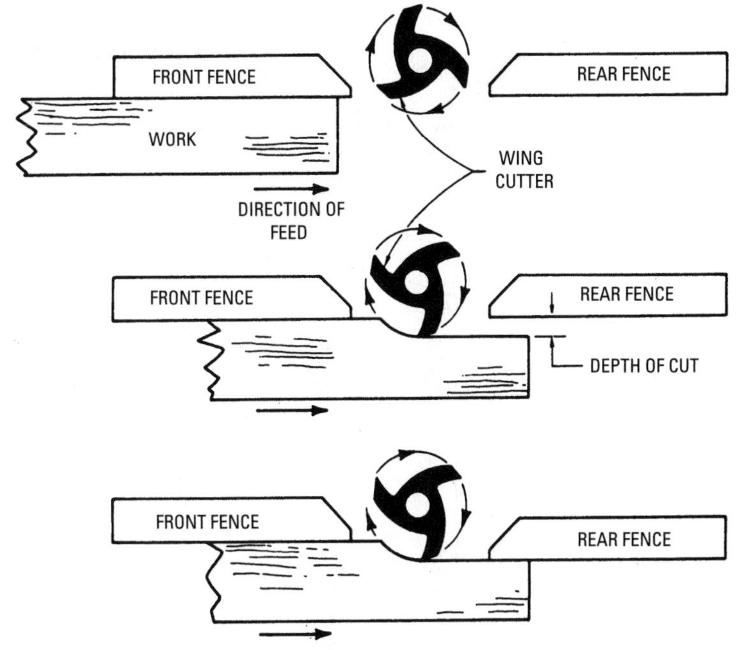

Figure 7-49 Adjustment of shaper fences.

When matching curved and irregularly shaped parts, some operators use small curved fences equipped with adjusting rods that fit into the regular guard and hold-down device. They usually find this method better than working the stock against the shaper collar, as the friction produced by the latter method frequently burns and glazes the stock (Figs. 7-50 through 7-52).

Power and Speed

The size of the motor necessary to operate a shaper depends upon the type of work to be done. The medium-sized shaper using $^1/_2$-inch hole cutters works satisfactorily with a $^1/_2$-horsepower motor. In cases where large knives mounted between slotted collars are used, a $^3/_4$- to 1-horsepower motor will be necessary.

The motor must be a 3450-rpm type in order to give the shaper spindle the required speed. With a pulley ratio of 3 to 1, the actual spindle speed will be in the order of 10,000 rpm. The motor should preferably be of the reversible type, since an opposite direction of rotation may often be required.

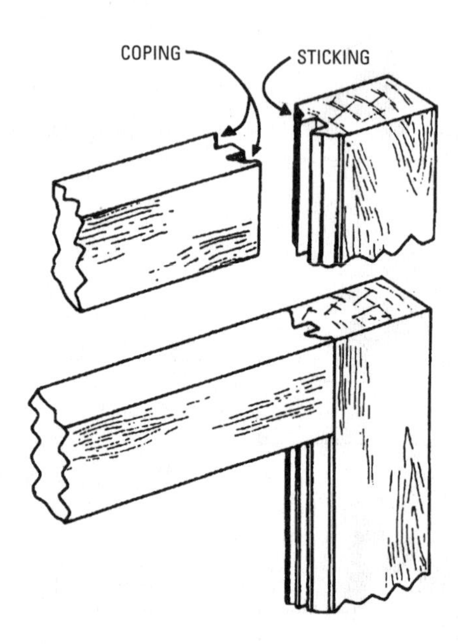

Figure 7-50 A shaper is used to form the parts of a door.

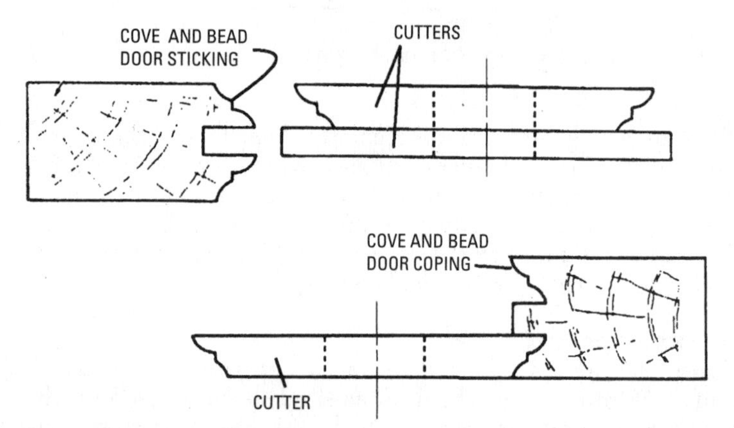

Figure 7-51 The same cutter is used to form both mating sections of a door.

Radial Arm Saw

Some craftsmen consider the *radial arm saw* to be the most important stationary power tool, even more important than the bench or table saw, which is favored by many. In fact, the radial arm saw (Figure 7-53) does all the things a table saw does. With the table saw,

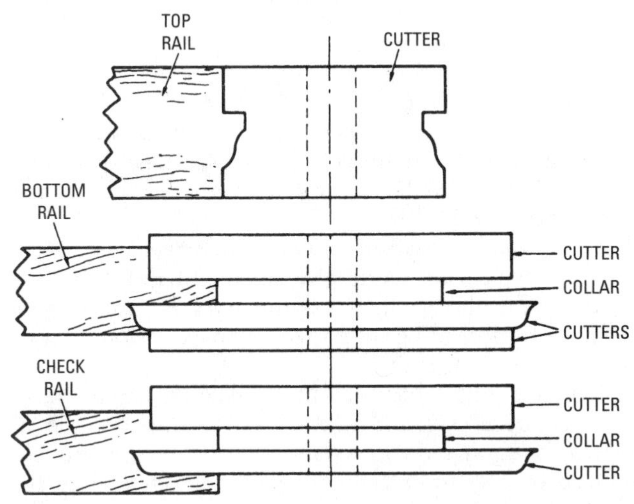

Figure 7-52 Shaper cutters used to form the parts of a window sash.

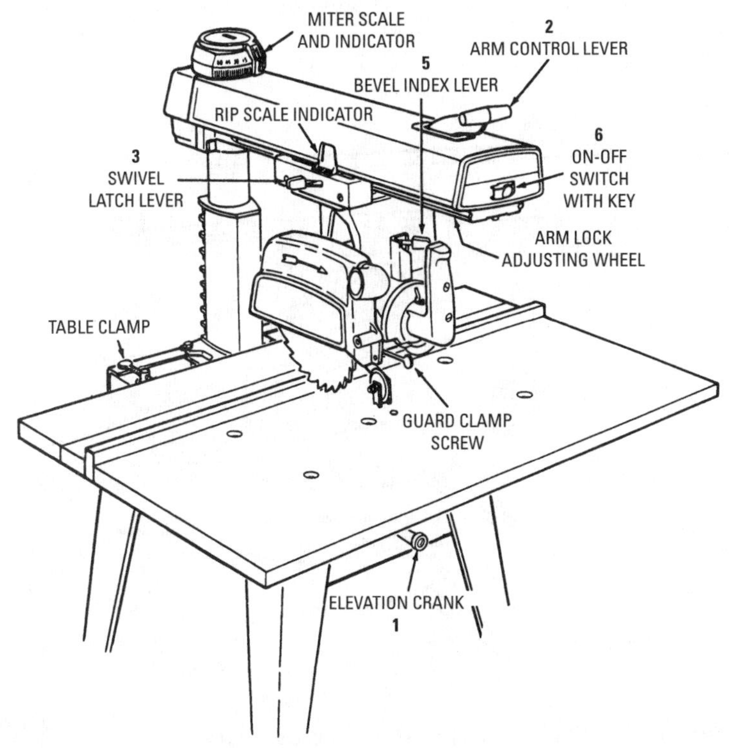

Figure 7-53 Radial arm saw.

however, you move the work into the revolving saw blade. With the radial arm saw, the saw, which is suspended on an arm, is moved into the work. For example, if you are doing a crosscutting operation, the work is set on the table, and then the saw is turned on and moved across the work, severing it neatly.

The radial arm saw is also capable of many other cuts. Most commonly there is the rip cut (cutting stock lengthwise). Here, the work is moved into the blade, which is stationary. It is also common practice for the workpiece to be supported on an extension table of some sort so that it is manageable.

The saw, complete with motor, can be tilted so that it can make miter cuts, crosscut bevel cuts, and others. Like the table saw, the radial arm saw may be equipped with a dado head for making grooves in stock. In this case, instead of feeding the work into the blade, the dado head is drawn across the stock.

The radial arm saw may be equipped with a variety of attachments that can be used for drilling holes, sanding, drum sanding, and buffing. Figs. 7-54 to 7-57 show a variety of operations that can be performed with the saw.

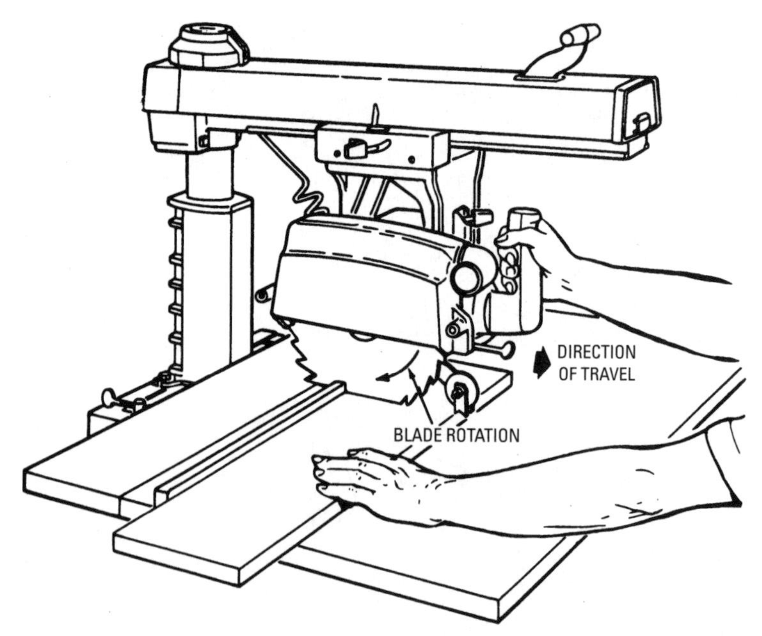

Figure 7-54 The basic operation is crosscutting.

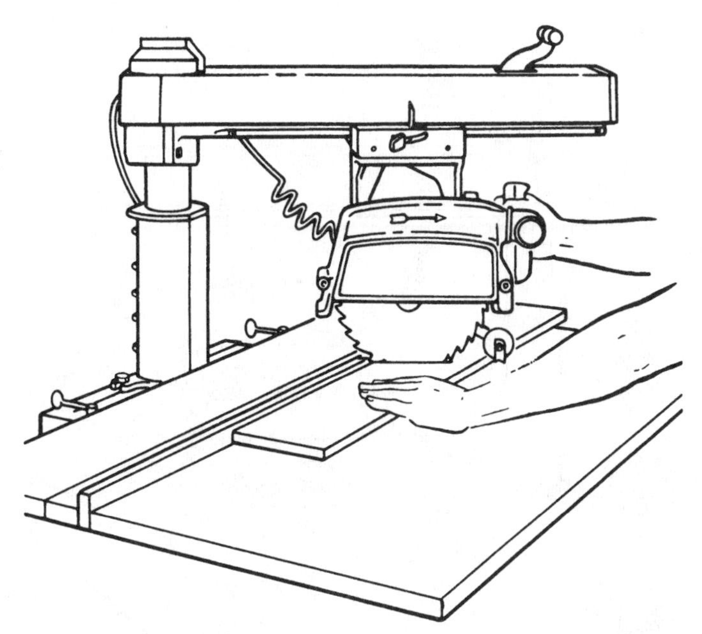

Figure 7-55 Miter crosscut. Saw is set to angle desired.

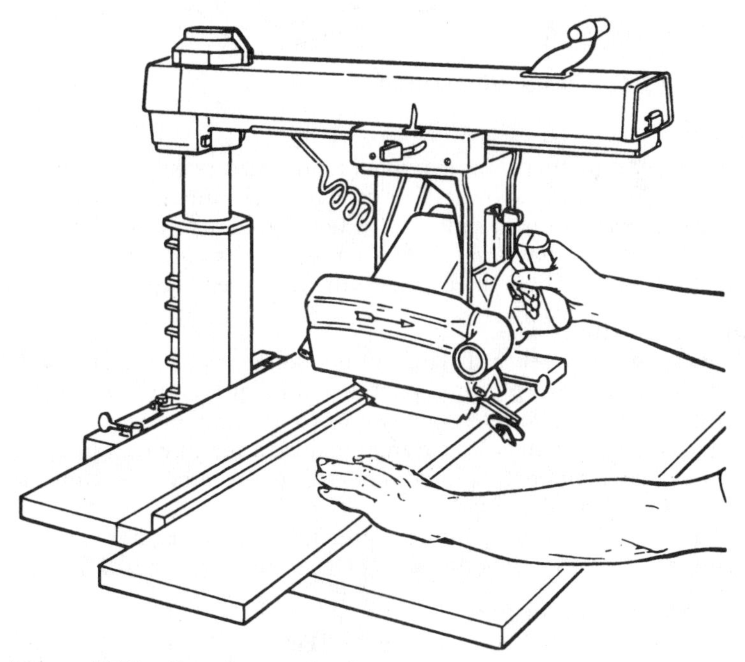

Figure 7-56 Bevel crosscut.

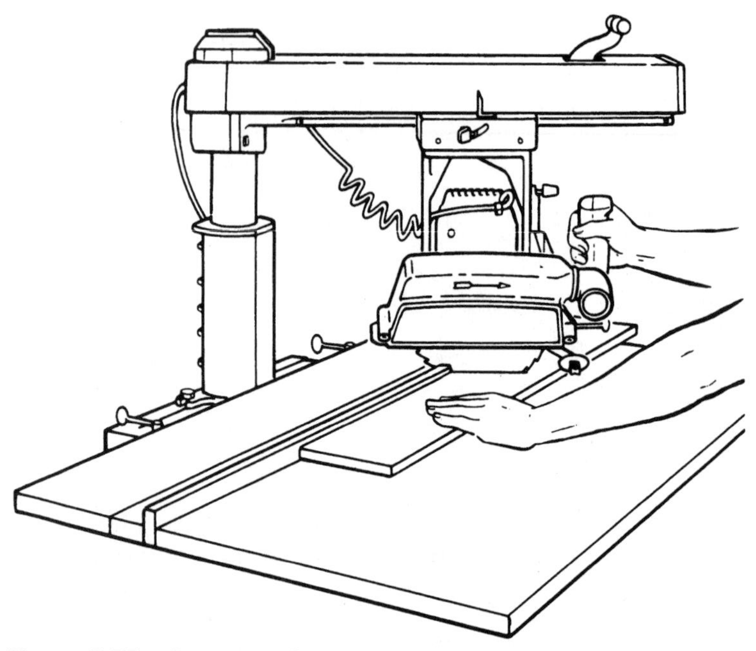

Figure 7-57 Compound crosscut.

Sander

Sanders play an important role in all kinds of carpentry and woodworking. You can sand by hand or use a portable sander. Alternately, you can use a stationary sander, of which there are four main types: the *drum sander*, *belt sander*, *disk sander*, and a combination belt and disk sander. These three sanders are discussed here.

Drum Sanders

Drum sanders are generally large production machines used to sand items such as doors, plywood panels, and other types of flat work. The drum sander is also frequently used to surface the faces of panel frames (after assembly of the frames on jigs) after gluing to smooth out any slight differences in adjoining frame members that might cause gaps in the glue bond.

The drum sander is built on the principle of a planer, but has one or more rotating sanding drums in place of the cutter heads.

Belt and Disk Sanders

The combination belt and disk sander (Figure 7-58) is used chiefly for cabinet work and other small stock. The disk table may be moved

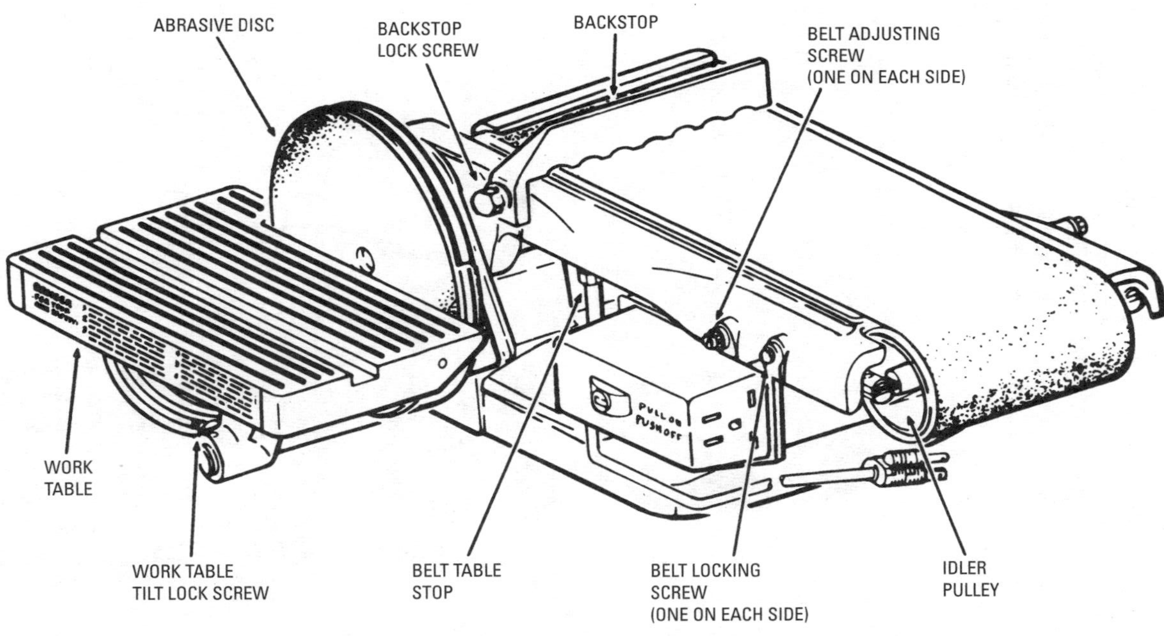

Figure 7-58 Belt-disk sander anatomy.

up and down and tilted to any angle from 0° to 45°. It is also fitted with gages for grinding compound angles, for making duplicate pieces, and for finishing circular pieces of various radii. The disk sander may be used for much circular work that is turned in the lathe, and for many operations for which the planer and jointer are used. A separate belt sander is shown in Figure 7-59 and a disk sander is shown in Figure 7-60. Figure 7-61 shows a typical use of a disk sander—sanding end grain.

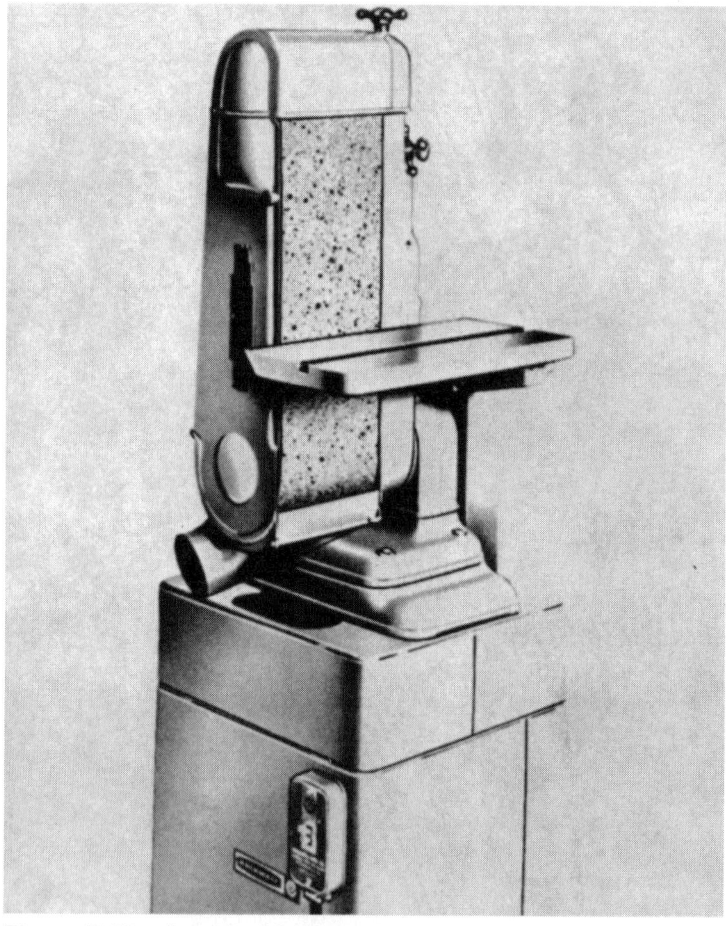

Figure 7-59 A typical belt sander.

Figure 7-60 Disk sander.

Figure 7-62 shows the use of the miter gage. By using this gage in combination with the tilting table, it is comparatively simple to obtain any combination of angles desired on the work.

A miter table is also provided for the sanding belt. By using this table and miter gage, angles can be sanded on the belt, as shown in Figure 7-62.

Drill Press

A *drill press* is really the only machine to consider for precision drilling of holes (Figure 7-63). Essentially, it consists of a drill that

Figure 7-61 A disk sander being used to sand the end grain of a piece of wood.

rides up and down, controlled by a feed lever, on a solid column. The workpiece is placed on a table (that can be raised or lowered as desired), and the hole is drilled. The size hole drilled depends on the bit used with the tool. The size of the drill press itself is determined by the diameter of the circle of the workpiece. The distance from the center of this piece to the column determines size. Hence, if the distance from the center of the table to the column is 10 inches, a 20-inch disk can be fitted on the table, and the tool is characterized as 20 inches.

The speed of the drill press is controlled by the rpm of the motor and the belt, which is looped over several pulleys.

The drill press is also capable of routing and shaping, but these are secondary rather than main functions. Its main function is to drill holes of various sizes with precision (Figure 7-64 and Figure 7-65).

The speed range of a drill press depends upon the size and number of pulleys, as well as the speed of the driving motor or shaft.

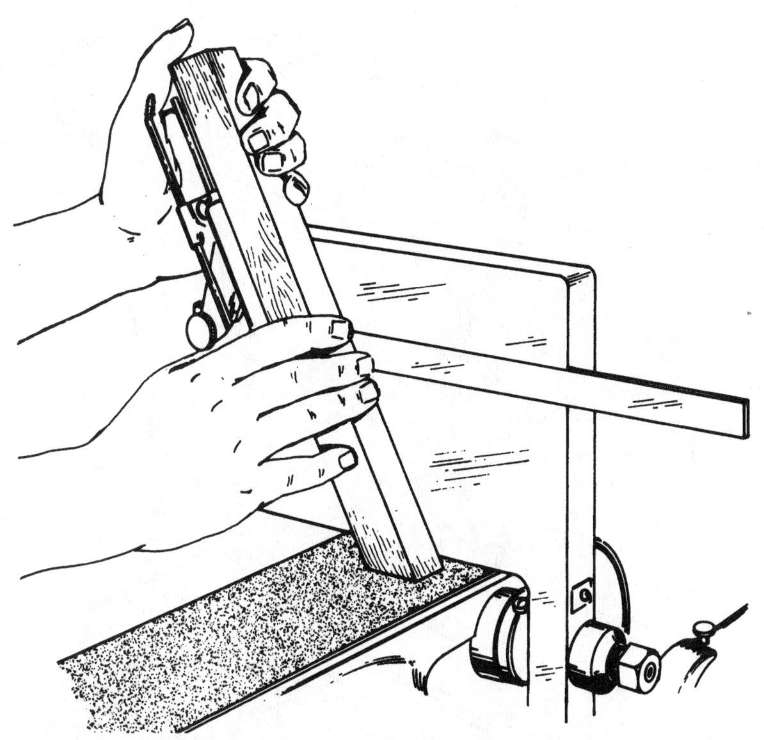

Figure 7-62 Use of the vertical table and miter gage to sand the end of a piece of wood. The table ensures an accurate surface that is 90° with relation to the sanding belt. The miter gage determines the accuracy of the bevel.

In a typical machine equipped with a four-step cone pulley, the speed may be varied from 600 to 5000 rpm. The speed of a machine used exclusively for metalworking ranges between 400 and 2000 rpm.

All cutting tools used for making holes in wood are called *bits*, whereas similar tools used for metals are called *drills*. A distinction is sometimes made in the cutting operation, the term *boring* applying to holes made in wood, while *drilling* means a cutting operation in metal. The most common type of machine boring bits are the *auger* and *twist-drill*. Other types are the *center* and *expansive bit*, each of which has its particular application in the woodworking shop.

Auger bits are equipped with a central screw that draws them into the wood, two ribs that score the circle, and two lips or cutters that

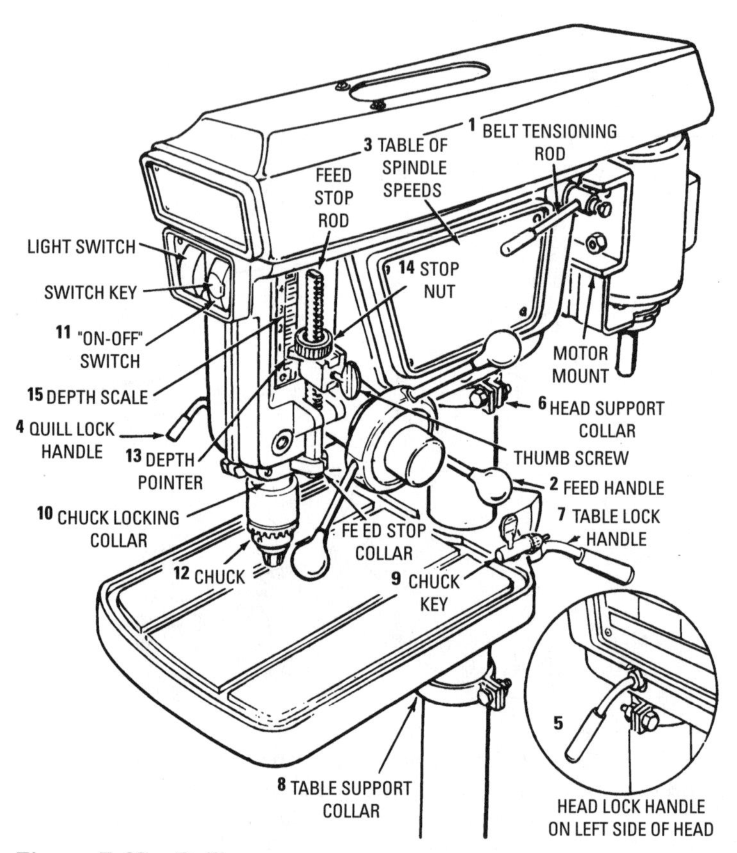

Figure 7-63 Drill press.

cut the shavings. These are used for boring holes from $^1/_2$ to 2 inches. The sizes are listed in 16ths; thus a 2-inch auger is listed as a number 32.

A twist drill differs from an auger in the absence of a screw and a less acute angle of the lip. Hence, there is no tendency to split the wood. In other words, the tool does not pull itself in by a taper screw, but enters by external pressure. Twist drills are used for boring small holes where the ordinary auger would probably split the wood. They come with either round or square shanks and in sizes from $1^1/_{16}$ to $^5/_8$ inch or larger, varying by 32nds.

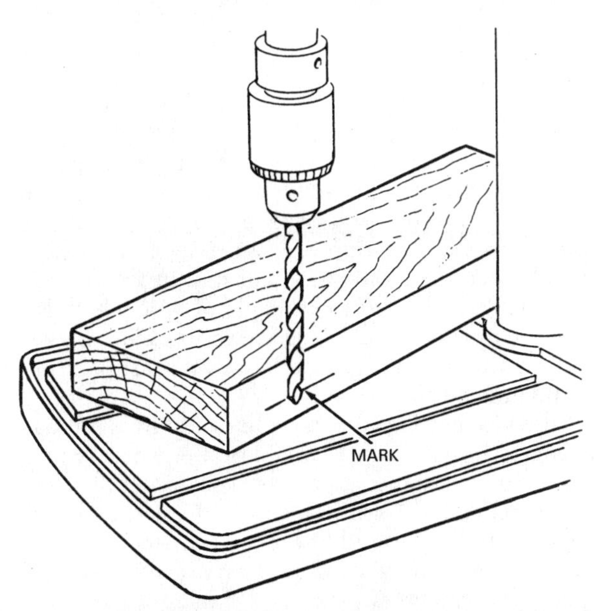

Figure 7-64 Blind drilling of holes is easy with a drill press. Just set the bit to go down a certain amount.

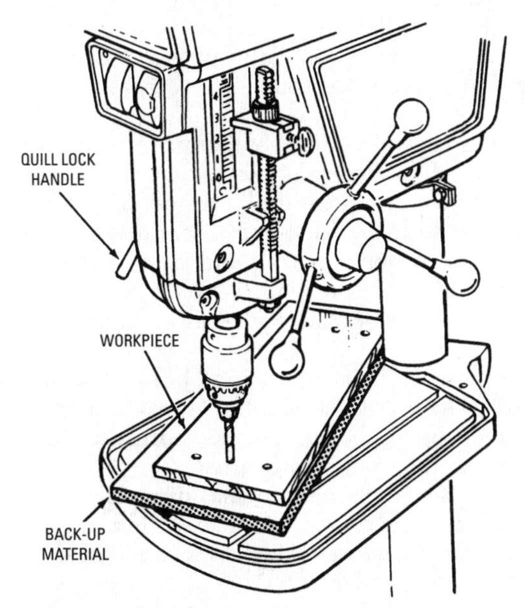

Figure 7-65 When drilling through a material, a back-up piece of material is a good idea. It avoids splintering on the far side of material.

Summary

The efficiency of a table saw depends mainly on the type and condition of the blade used. There are generally three types of table-saw blades: crosscut, ripsaw, and combination. Many operations can be performed on a table saw, such as ripping, rabbeting, dado cutting, miter cutting, molding cutting, and tenon cutting. In the conventional design, the saw can be tilted to an angle of 45°, which is a safety factor since the operator does not have to work in an awkward position. When wood is cut on a table saw, it must be firmly held against a metal guide or fence.

Band saws are used for cutting curved outlines and lines not parallel to a straight edge. The type most adaptable to the general woodworking shop use is called a band scroll saw. They generally use a blade 1/2 to 1 inch in width and 10 to 40 inches in diameter, depending on the diameter of the upper and lower drive wheel. Although a band saw is excellent for curved cutting, it may also be used for making straight cuts for both crosscutting and ripping when a table saw is not available. When using a band saw for straight cutting, it is easier to follow a straight line with a wide blade.

The jigsaw and band saw generally do the same type of work, although the jigsaw is more adaptable to sharp curves, because smaller and finer blades are used. The jigsaw can cut inside holes since the blade is easily removed and inserted through a hole bored in the wood.

The speed of a jigsaw is generally varied by selecting the proper pulley groove and is governed by the material used and the type of work to be done. The blades vary a great deal in length, thickness, width, and number of teeth per inch. Blades are generally known as saber blades (which connect at only one end) and jeweler's blades (which connect at both ends). Since the blade action is an up-and-down motion, sharper corners can be made and more accurate work can be accomplished.

The principal parts of a wood lathe are the bed, headstock, tailstock, and the tool post. The headstock and the tailstock support the wood, which is being turned. The tool post supports the chisel or gouge that cuts or shapes the wood being turned.

There are many woodturning tools, both for cutting and for measuring. Cutting tools include skew chisels, spear-point chisels, parting tools, round-nose chisels, and gouges. The skew chisel is beveled on both sides and forms a cutting edge of 60°.

A planer is a stationary tool used to plane wood by means of removing the rough surface. The wood to be planed is passed under the cutting heads, leaving a smooth or finished surface. Many machines are available with two cutting heads so that both sides of a piece of lumber may be planed in one operation. A planer is generally power-fed, which means the lumber is automatically fed into the cutting heads.

The jointer is very similar to the planer, except that it has only one head generally located below the table. The lumber is usually fed by hand. The primary use of the jointer is to cut a true surface on lumber that is warped or twisted.

A shaper is used to finish the edge on stock lumber, and for rabbeting, grooving, and fluting. Many times, stock may be cut roughly to shape with a band saw or jigsaw before it is finished on the shaper.

The radial arm saw does all the things a table saw does. With the table saw, however, you move the work into the revolving saw blade. With the radial arm saw, the saw, which is suspended on an arm, is moved into the work. The saw, complete with motor, can be tilted so that it can make miter cuts, crosscut bevel cuts, and others.

The function of a sanding machine is to smooth the wood surface by a scratching action of various types of abrasives. These abrasives come in a wide variety of sizes, shapes, and coarseness. There are various types of sanding machines, such as drum, belt, and disk. Drum sanders are generally large production machines used to sand items such as doors, plywood panels, and other large flat work. Certain drum sanders are designed to sand or finish both top and bottom surfaces of a board simultaneously.

A combination belt and disk sander is used mainly for cabinet work and other small jobs. The disk sander is used for circular work and various angles. Generally, the sanding table can be moved up and down or tilted to any angle up to 45°. Gages are also used to grind or sand compound angles or for making duplicate pieces.

Essentially a drill press consists of a vertical spindle having a chuck at one end and a telescoping splined sleeve with a drive pulley on the other. A table is placed under or in front of the chuck to hold the work. The size of the boring machine depends on the size of work to be bored and the speed of the drill. The cutting tools used for making holes in wood are called bits.

Review Questions

1. Name the various types of sanding machines.
2. Name the sanding machine used for cabinet work or for various angles.

3. Explain the difference between a planer and jointer.
4. What are the principal operations of a shaper?
5. How many cutting heads are used on a planer?
6. Name the four main parts of the cutting portion of a wood lathe.
7. What is the difference between inside calipers and outside calipers?
8. What is the difference between dividers and calipers?
9. What is the motion action of the jigsaw blade?
10. What are some of the advantages of the jigsaw over the band saw?
11. Name the two types of blades used in the jigsaw.
12. What is the throat opening on a jigsaw?
13. How is the speed change on a jigsaw accomplished?
14. What is the big advantage in using a band saw?
15. How many teeth per inch are there on the common band-saw blade?
16. How should tension on the saw blade be checked?
17. How is straight cutting accomplished?
18. How is miter cutting accomplished?
19. Why is it better to tilt the saw blade than to tilt the table in miter cuts?
20. Explain the purpose of the fence.
21. What is a dado head?
22. What is rabbeting?
23. What is a drill?
24. What is a bit?
25. Explain the purpose of a drill chuck.
26. What is the difference between an auger bit and an expansion bit?

Chapter 8

Insulation and Energy Conservation

Ever since the energy crunch began, there has been a tremendous interest in energy saving on new work and existing structures. Carpenters, contractors, builders, and do-it-yourselfers have found that, over time, using certain materials and techniques can result in big savings.

What follows is a roundup of energy savers for new and existing buildings. In most cases, the suggestions can be readily adopted. However, it is always a good idea to have discussions with knowledgeable people (such as experts at local utility companies) to see what works best. For example, 12-inch-thick insulation may be just the thing for frigid New Hampshire winters, but it would hardly be practical (in terms of original cost) for insulating a stucco home south of Los Angeles.

Insulation—R Factors

A key concept to remember when selecting insulation is the *R factor* of the material. *R* stands for resistance: the ability of the material (technically the stratified air spaces) to resist or retard heat. In the winter, this resistance works by keeping warm air inside the home. In the summer, it keeps hot air out.

The critical time is winter, when the expensive fuel used to heat that warm air may be traveling out of the house by conduction, making the furnace work harder and using up more fuel.

Cold air infiltration is also an energy factor. If the house has cracks and openings where cold air can sneak in and warm air escape, this will result in higher fuel costs. Such openings must be plugged up.

Insulation comes in a variety of types, generally ranging from about R-4 to R-38. All insulation has its R factor stamped on it. The R-4 material could be urethane boards applied to basement walls, the R-38 material could be 12-inch-thick fiberglass batts. The following sections discuss the materials available:

Rigid Board Insulation

This is available in various sizes in board form. It may be urethane, fiberglass, Styrofoam, or composition board (Figure 8-1). It is easily worked with standard hand tools, and some contractors favor it as sheathing. While the R values are not high compared with other

Figure 8-1 A block or masonry wall is a good candidate for Styrofoam or urethane board insulation. *(Courtesy of Dow)*

kinds of insulation, they are better than those for board or plywood. Rigid board insulation is also handy to use in the basement, where it can be glued directly to walls. Before using it, however, you should check for fire restrictions. Some rigid board insulation is flammable and must be covered by a nonflammable material before any other flammable material, such as wood paneling, can be installed (Figure 8-2).

Siding with Insulation

Some siding (chiefly aluminum) has built-in insulation. R factors vary, but they should be checked out before any material is purchased.

Batts and Blankets

Batts are flexible insulation sections cut to fit between standard studs (on 16-inch centers) that are about 8 feet long. *Blankets* are the same width as batts, but come in many lengths up to 100 feet. Batts and blankets may be faced on both sides with a paper-like material, and one side will have a vapor barrier (either aluminum

Figure 8-2 Material is glued on. Gypsum board must be used over plastic if material is combustible. *(Courtesy of Certain-Teed)*

or Kraft-paper), or may be unfaced (Figure 8-3). Mineral fiber is the filler for such insulation: fiberglass (such as that made by Owens-Corning) or mineral wool. On the edges of batts and blankets are strips that are stapled to framing members.

Batts and blankets are favored by contractors and do-it-yourselfers alike. This insulation is used on exposed stud walls, on attic floors (between joists), and on open ceilings. It ranges in R factor from around R-4 to R-38.

Fill-Type Insulation

Another insulation is the fill type (Figure 8-4), which comes as a granular material (Vermiculite) or a macerated paper (cellulose). This material may be poured in place (such as between attic joists), but it is primarily blown into areas inaccessible for installing other

Figure 8-3 In the attic, blanket insulation fits between joists. Use plywood as temporary floor during installation. *(Courtesy of Owens-Corning)*

Figure 8-4 Fill insulation (mineral wool) being blown into attic. *(Courtesy of Mineral Wool Mfg. Assn)*

insulation, mainly walls in finished homes. Sections of siding are removed, holes are drilled between pairs of studs, a hose is inserted, and the material is blown into place. It fills up the cavities and insulates.

Foam

Like fill-type insulation, foam is used on finished work that is inaccessible to the batts or blankets. The foam is mixed and is applied as a liquid. It becomes foam, filling the spaces. When it cures it becomes hard. There has been controversy about foam insulation. Complaints have been voiced (and indeed the government has investigated them) about medical problems caused by the fumes of foam, and its use has been banned in many places. In fairness to foam makers and installers, though, a wise precaution is to know the type of foam an installer plans to use. Foams are considered harmless. However, they may suffer from a blanket indictment of all foam. Foam is practical because nothing has a higher R value, inch-for-inch, than foam.

Reducing Leakage Around Doors and Windows

A large amount of the heat delivered to rooms is lost by air leakage around doors and windows. Another factor in heat loss is heat leakage through window or door glass, or through exterior doors.

These various heat losses may be greatly prevented by

- Caulking
- Weather strips
- Storm or double-glass windows
- Storm doors

Though the house may have been tight when built, a few years of usage will sometimes permit a considerable amount of leakage. In addition, small cracks often develop in masonry walls that, in time, may become sufficiently large to permit a considerable air leakage.

Caulking

The most common remedy for air leakage is found in caulking or plugging the leaks. A variety of caulks is available, from oil-base to butyl rubber. Caulk is easiest to use in cartridges that are slipped into a caulking gun (Figure 8-5). The gun extrudes a thin bead of

Figure 8-5 Caulk is important in the fight to save energy. *(Courtesy of United Gilsonite)*

caulk, which will effectively seal off cracks. Generally, cracks form where dissimilar materials meet, such as on wood trim and siding. For deep cracks, oakum or mineral fiber can be packed into the crack before the caulk is used.

Weatherstripping

Weatherstripping is another excellent idea. Weatherstrip comes in a variety of types and is designed to be used where there is a natural opening in the house (such as beneath the garage or exterior doors, or where windows meet framework). You can use felt material, which can be tacked in place to V-shaped strips that have one adhesive edge. Also available is a moldable weatherstrip and adhesive-backed felt. Whichever kind you use, weatherstripping is an energy saver.

Window and Door Glazing

A large amount of heat can be lost through window glass. Thus storm windows and storm doors are an excellent idea. Indeed, they are just as important as insulation. If you find that you cannot afford storm windows this winter, you can use polyethylene sheeting and make your own. Polyethylene is as effective as glass, though it is only temporary, of course.

An even better solution for windows is double- and even triple-glazed windows. A dead air space is trapped between the layers of glass and retards heat transmission. Such glazing is initially expensive, but it pays for itself over time.

Where to Insulate

A basic question, of course, is where to insulate. Generally, all walls, ceilings, roofs, and floors that separate heated from unheated spaces should be insulated. Insulation should be placed on all outside walls and in the ceiling, as shown in Figure 8-6. In houses with unheated crawl spaces, insulation should be placed between the floor joists. If a blanket type of insulation is used, it should be well-supported with galvanized mesh wire or by a rigid board. The most important area to insulate, though, is the attic, simply because heat rises. The greatest amount of heat is lost through the attic.

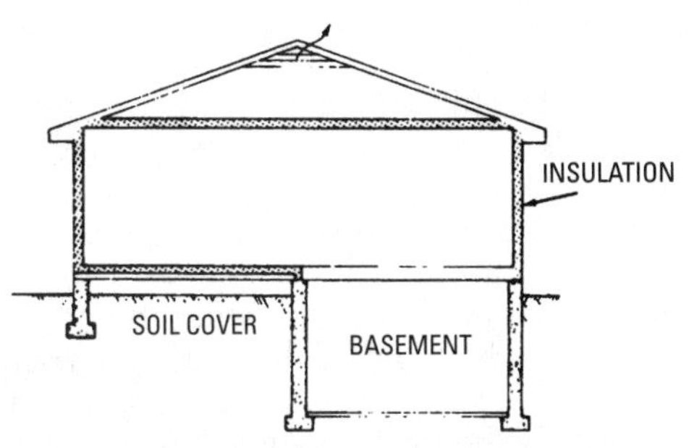

Figure 8-6 Areas to be insulated. Attic or top-floor ceiling is most important.

In $1^1/_2$-story houses, insulation should be placed along all areas that are adjacent to unheated areas, as shown in Figure 8-7. These include stairways, dwarf walls, and dormers. Provisions should be made for ventilation of the unheated areas. Where attic storage space is unheated and a stairway is included, insulation should be used around the stairway as well as in the first floor ceiling, as shown in

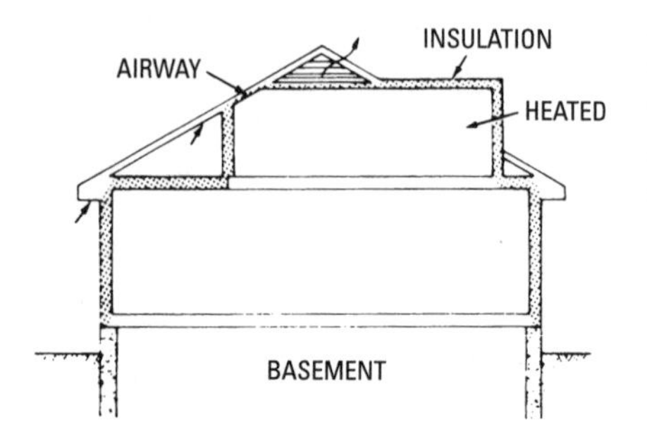

Figure 8-7 Method of insulating 1 $^{1}/_{2}$-story house. Basement wall could also be insulated with rigid board.

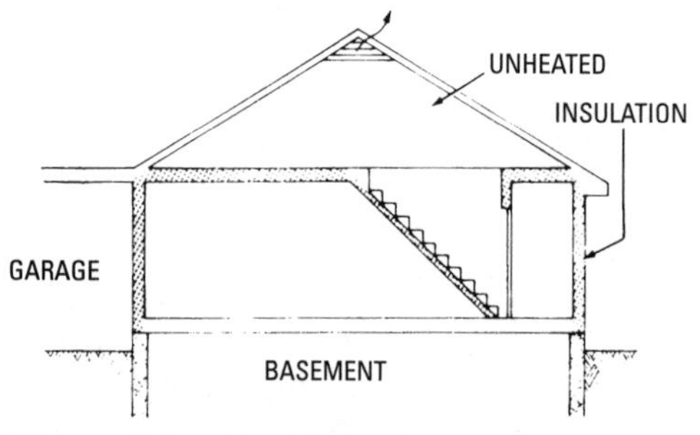

Figure 8-8 Method of shielding a heated area from an unheated area by insulating.

Figure 8-8. The door leading to the attic should be weatherstripped to prevent the loss of heat. Walls adjoining an unheated garage or porch should be well-insulated.

In houses with flat or low-pitched roofs (Figure 8-9), insulation should be used in the ceiling area with sufficient space allowed above for a clear ventilating area between the joists. Insulation should be used along the perimeter of a house built on a slab. A vapor barrier should be included under the slab. Insulation, as mentioned, can be used to improve comfort within the house during hot weather. Surfaces exposed to the direct rays of the sun may attain temperatures

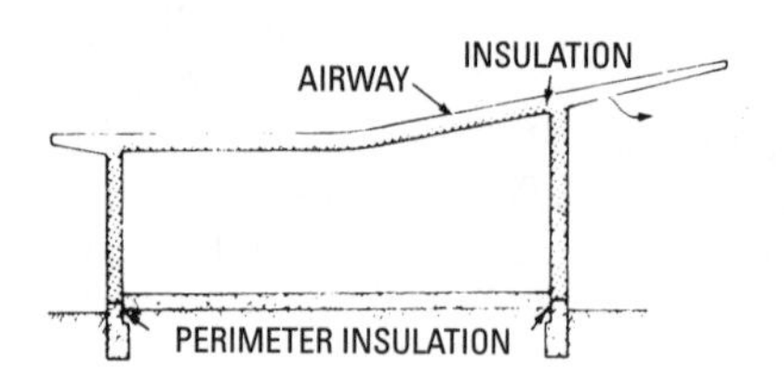

Figure 8-9 Insulation of a typical flat-roof house.

of 50°F or more above shade temperature and, of course, tend to transfer this heat toward the inside of the house. Insulation in roofs and walls retards the flow of heat. In addition, less heat is transmitted through such surfaces.

Where any cooling system is used, insulation should be used in all exposed ceilings and walls in the same manner as for preventing heat loss in cold weather. Of course, where air conditioning is used, the windows and doors should be kept closed during periods when outdoor temperatures are above inside temperatures. Windows exposed to the sun should be shaded with awnings.

Ventilation of attic and roof space is an important adjunct to insulation. Without ventilation, an attic space may become very hot and hold the heat for many hours. Obviously, more heat will be transmitted through the ceiling when the attic temperature is 150°F than when it is 120°F. Ventilation methods suggested for protection against cold-weather condensation apply equally well to protection against excessive hot-weather roof temperature. A detailed discussion of ventilation appears later in this chapter.

Installation of Insulation

Blanket or batt insulation should be installed between framing members so that the tabs lap all the face of the framing members, or on the sides. The vapor barrier should face the inside of the house. Where there is no vapor barrier on the insulation, a vapor barrier should be used over these unprotected areas (Figure 8-10). A hand-stapling machine is usually used to fasten insulation tabs and vapor barriers in place. A vapor barrier should be used on the warm side (the bottom, in case of ceiling joists) before insulation is placed. Figs. 8-11 through 8-17 show a number of other installation tips.

Figure 8-10 The right way to install a vapor barrier.

Vapor Barriers

Most building materials are permeable to water vapor. In cold climates during cold weather, such vapor (generated in the house from cooking, dishwashing, laundering, bathing, humidifiers, and other sources) may pass through wall and ceiling materials and condense in the wall or attic space. There it may subsequently do damage to

Figure 8-11 **Installation in attic floor. Start with installation of proper vent to soffitt space under eaves. These vents are made on-site of corrugated cardboard and stapled onto roof sheathing. Next, tuck fiberglass blankets against vents between rafters and joists.** *(Courtesy of Owens-Corning)*

Figure 8-12 **Light fixture. To prevent fire, keep insulation at least 3 inches away from recessed light fixtures unless the fixture is rated "I.C." for direct contact of insulation in an insulated ceiling.** *(Courtesy of Owens-Corning)*

Figure 8-13 Additional insulation is added by laying a second layer of blankets or batts perpendicular to the first layer. *(Courtesy of Owens-Corning)*

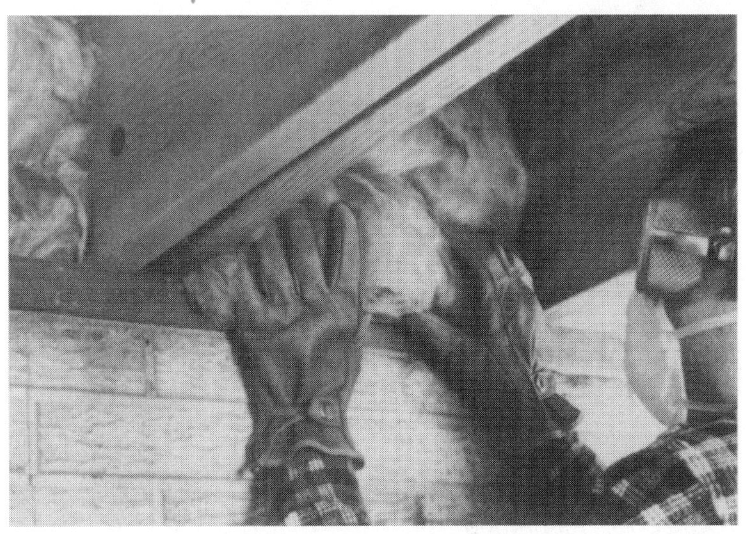

Figure 8-14 In the basement, tuck insulation above the sill between and along joists. *(Courtesy of Owens-Corning)*

Figure 8-15 Fit blankets between floor joists and hold in place with flexible wire rods. *(Courtesy of Owens-Corning)*

Figure 8-16 Floors above unheated spaces should be insulated. Wire netting can be used to retain material.

Figure 8-17 A product from Du Pont called Tyvek is being used to cut air infiltration in an attic. The product is stapled into place.

exterior paint and interior finish, or may even cause decay in structural members. As a protection, material highly resistant to vapor transmission is called a *vapor barrier.* It should be used on the warm side of any unfaced insulation, as mentioned earlier. The most widely used vapor barrier material is polyethylene sheeting. This is sold by the foot off rolls and can be attached easily.

Some batt and blanket insulations have a vapor barrier on one side of the blanket. Such material should be attached on the sides or faces of the studs. Where polyethylene is used, it should be applied vertically over the face of the studs and stapled on. The wall material is then applied over the vapor barrier. Vapor barriers should be cut to fit tightly around electric outlets and switch boxes. Any cold-air returns to the furnace in outside walls should be lined with tin.

Ventilation

It is an excellent idea to button up a house with insulation and other energy-saving steps. However, it is also important to ventilate a house properly to avoid condensation and high temperatures.

The attic and crawl spaces, in particular, deserve attention. Vents (power types and plain standard vents) let moisture escape, and in summer, the moving air in an attic tends to keep it cooler. Indeed, an attic can reach an almost unbelievable 140°F in summer.

Attics

At least two vent openings should always be provided. They should be located so that air can flow in one vent and out the other. A combination of vents at the eaves and gable vents is better than gable vents alone. A combination of eave vents and continuous ridge venting is best (Figure 8-18).

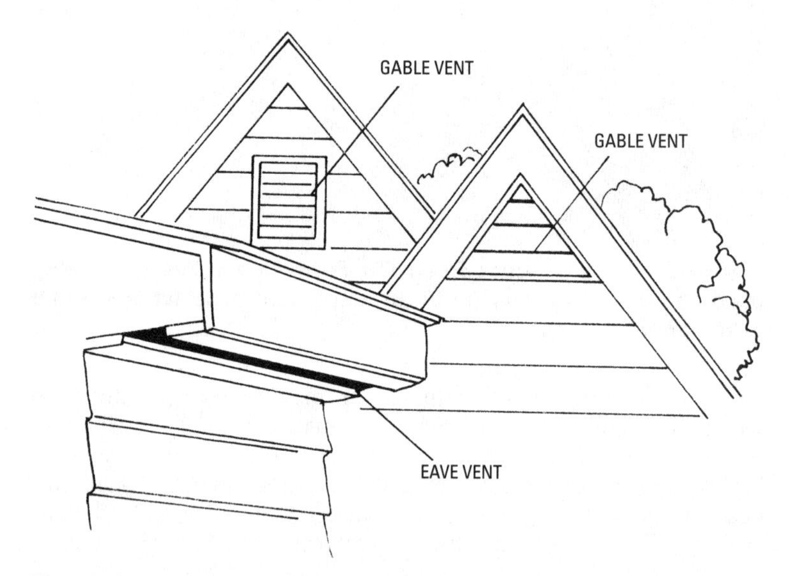

Figure 8-18 Vent arrangements.

An attic should have a minimum amount of vent area, as follows:

- *Combinations of eave vents and gable vents without a vapor barrier*—1 square foot inlet and 1 square foot outlet for each

600 square feet of ceiling area with at least half the vent area at the tops of the gables and the balance at the eaves

- *Gable vents only without a vapor barrier*—1 square foot inlet and 1 square foot outlet for each 300 square feet of ceiling area
- *Gable vents only with a vapor barrier*—1 square foot inlet and 1 square foot outlet for each 600 square feet of ceiling area

Crawl Spaces

For a crawl space that is unheated, there should be at least two vents opposite one another. The minimum opening size, with moisture seal (4-mil or thicker polyethylene sheeting or 55-lb asphalt roll roofing lapped 3 inches) on the ground should be 1 square foot for every 1500 feet of crawl space.

If there is no moisture seal, then there should be an opening of 1 square foot for every 150 square feet.

Covered Spaces

When attic or crawl space vents are covered by screening or other materials, this results in a reduction in the size of the opening. Opening sizes should be increased, as indicated in Table 8-1.

Table 8-1 Opening Sizes

Covering	*Size of Opening*
1/4-inch hardware cloth	1 × net vent area
1/4-inch hardware cloth and rain louvers	2 × net vent area
1/8-inch screen	1 1/4 × net vent area
1/8-inch screen and rain louvers	2 1/4 × net vent area
1/16-inch screen	2 × net vent area
1/16-inch screen and rain louvers	3 × net vent area

Figure 8-19 shows an example of an energy-conscious home.

Building Modifications

To make housing more efficient, it is necessary to make some modifications in present-day carpentry practices. For example, the following must be done to make room for better and more efficient insulation:

- Truss rafters are modified to permit stacking of two 6-inch thick batts of insulation over the wall plate. The truss is hipped by adding vertical members at each end and directly over the 2×6 studs of the outer wall.

- All outside walls use thicker (2×6) studs on 2-foot centers to accommodate the thicker insulation batts.
- Ductwork is framed into living space to reduce heat loss as warmed air passes through the ducts to the rooms.
- Wiring is rerouted along the sole plate and through notches in the 2×6 studs. This leaves the insulation cavity in the wall free from obstructions.
- Partitions join outside walls without creating a gap in the insulation. Drywall passes between the sole plates and abutting studs of the interior walls. The cavity is fully insulated.
- Window area is reduced to 8 percent of the living area.
- Box headers over the door and window openings receive insulation. In present practice, the space is filled with 2×10 boards or whatever is needed. Window header space can be filled with insulation. This reduces the heat loss through the wood that would normally be located here.

Summary

Insulation is a material used to prevent the high outside temperature from entering the house during the summer and the inside higher temperature from leaving the building during the winter.

The primary reasons for the application of insulating materials in homes are economy and comfort. When insulation is properly installed, it keeps the interior of the house comfortable with a minimum of fuel consumption. In the summer months, insulation helps keep rooms in all parts of the house comparatively cool and comfortable. If air conditioning is used during the hot season, the system may be operated with more economy if the building is properly insulated.

There are several insulating materials. Among the more common materials used for insulation in homes are batts and blankets.

Fill-type insulation materials are powdered, granulated, or shredded, and come in bulk lots. This type of insulation can be applied in houses under construction, or in those already completed, by pouring or blowing it into the space between the framing members. Fill insulation is generally placed between studs, ceiling joists, and attic joists.

The best insulation and its application will be of little help if air leakage around doors and windows is not controlled. The most common remedy against this air leakage is found in caulking or filling the leaks. Many caulking compounds are manufactured for various building materials.

Review Questions

1. Name the various insulating materials.
2. How is fill insulation applied in a house?
3. What is a vapor barrier?
4. What is the purpose of insulation?
5. What is used to control air leaks?
6. Why is the use of foam considered controversial?
7. What is the reason for wrapping a house in Tyvek?
8. What does the "R" stand for in the R value rating?
9. What material, used for siding, has built-in insulation?
10. What part or parts of a house should be insulated?

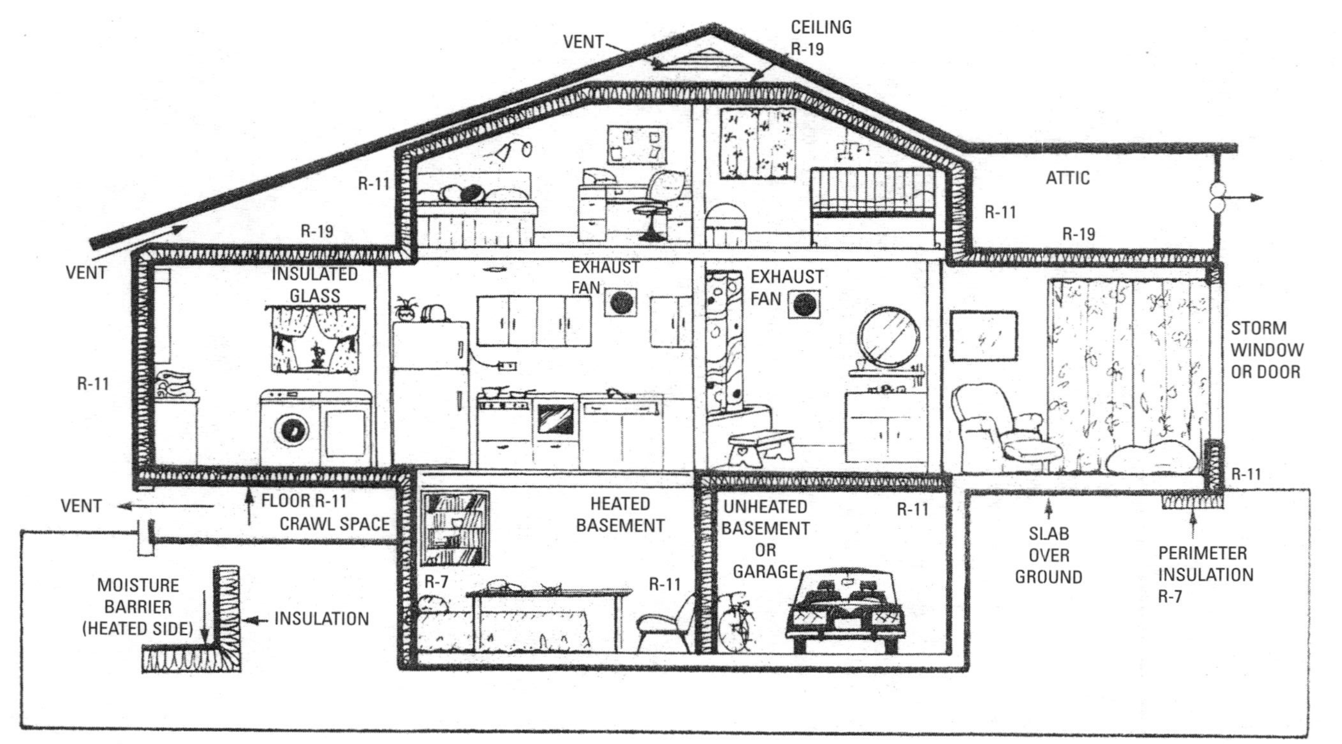

Figure 8-19 An energy-conscious home. *(Courtesy of Owens-Corning)*

Chapter 9

Solar Heat

Solar heating and cooling systems for the house are often thought of as a means of getting something for nothing. After all, the sun is free. All you have to do is devise a system to collect all this energy and channel it where you want it to heat the inside of the house or to cool it in the summer.

Sounds simple, doesn't it? It can be done, but it is can be expensive.

There are two ways to classify solar heating systems. The *passive* type is the simplest. It relies entirely on the movement of a liquid or air by means of the sun's energy. You actually use the sun's energy as a method of heating when you open the curtains on the sunny side of the house during the cold months. In the summer, you can use curtains to block the sun's rays and try to keep the room cool. The passive system has no moving parts. In the example, the moving of the curtains back and forth was an exception. The design of the house can have a lot to do with this type of heating and cooling. It takes into consideration whether the climate requires the house to have an overhang to shade the windows, or no overhang so that the sun can reach the windows and inside the house during the winter months.

The other type of system is called *active*. It has moving parts to add to the circulation of the heat by way of pumps to push hot water around the system or fans to blow the heated air and cause it to circulate.

Solar heating has long captivated people who want a free system to heat and cool their residences. However, as you examine this chapter, you will find that there is no free source of energy. Some types of solar energy systems are rather expensive to install and maintain. In this chapter you will learn

- How active and passive systems work
- How cooling and heating are accomplished
- How various solar energy systems compare with the conventional methods used for heating and cooling

Passive Solar Heating

Three concepts are used in the passive heating systems: indirect, direct, and isolated gain. Each of these concepts involves the relationship between the sun, storage mass, and living space.

Indirect Gain

In *indirect gain*, a storage mass is used to collect and store heat. The storage mass intercedes between the sun and the living space. The three types of indirect-gain solar buildings are mass trombe, water trombe, and roof pond.

Mass-trombe buildings involve only a large glass collector area with a storage mass directly behind it. There may be a variety of interpretations of this concept. One of the disadvantages is the large 15-inch-thick concrete mass constructed to absorb the heat during the day (Figure 9-1). The concrete has a black coating to aid the absorption of heat. Decorating around this gets to be a challenge. If it is a two-story house, most of the heat will go up the stairwell and remain in the upper bedrooms. Distribution of air by natural convection is viable with this system since the volume of air in the space between the glazing and storage mass is being heated to high temperatures and is constantly trying to move to other areas within the house. The mass can also be made of adobe, stone, or composites of brick, block, and sand.

Cooling is accomplished by allowing the 6-inch space between the mass and the glazed wall to be vented to the outside. Small fans may be necessary to move the hot air. Venting the hot air causes cooler air to be drawn through the house. This will produce some cooling during the summer. The massive wall and ground floor slab also maintains cooler daytime temperatures. Trees can be used for shade during the summer. They drop their leaves during the winter to allow for direct heating of the mass.

A hot-air furnace with ducts built into the wall is used for supplemental heat. Its performance evaluation is roughly 75 percent passive heating contribution. Performance is rated as excellent. For summer, larger vents are needed. In the winter, too much heat rises up the open stairwell.

Water-trombe buildings are another of the indirect-gain passive heating types. The buildings have large glazed areas and adjacent massive heat storage. The storage is in water or another liquid, held in a variety of containers, each with a different heat exchanger surface-to-storage mass ratio. Larger storage volumes provide greater and longer-term heat storage capacity. Smaller contained volumes provide greater heat exchange surfaces and faster distribution. The trade-off between heat exchange surface versus storage mass has not been fully developed. A number of different types of storage containers (such as tin cans, bottles, tubes, bins, barrels, drums, bags, and complete walls filled with water) have been used in experiments.

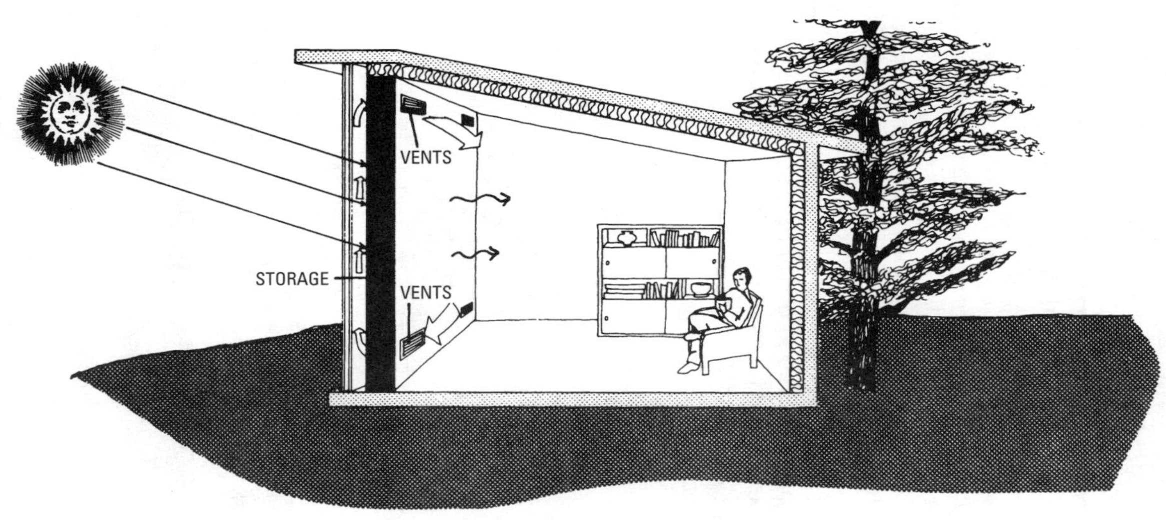

Figure 9-1 Mass-trombe building.

A gas-fired hot water heating system is used for backup purposes. That is primarily because this system has a 30 percent passive heating contribution. When fans are used to force air past the wall to improve the heat circulation, it is classified as a *hybrid system*.

The *roof-pond* type of building is exactly what its name implies. The roof is flooded with water (Figure 9-2). It is protected and controlled by exterior movable insulation. The water is exposed to direct solar gain that causes it to absorb and store heat. Since the heat source is on the roof, it radiates heat from the ceiling to the living space below. Heat is by radiation only. The ceiling height makes a difference to the individual being warmed, since radiation density drops off with distance. The storage mass should be uniformly spaced so it covers the entire living area. A hybrid of the passive type must be devised if it is to be more efficient. A movable insulation must be utilized on sunless winter days and nights to prevent unwanted heat losses to the outside. It is also needed for unwanted heat gain in the summer.

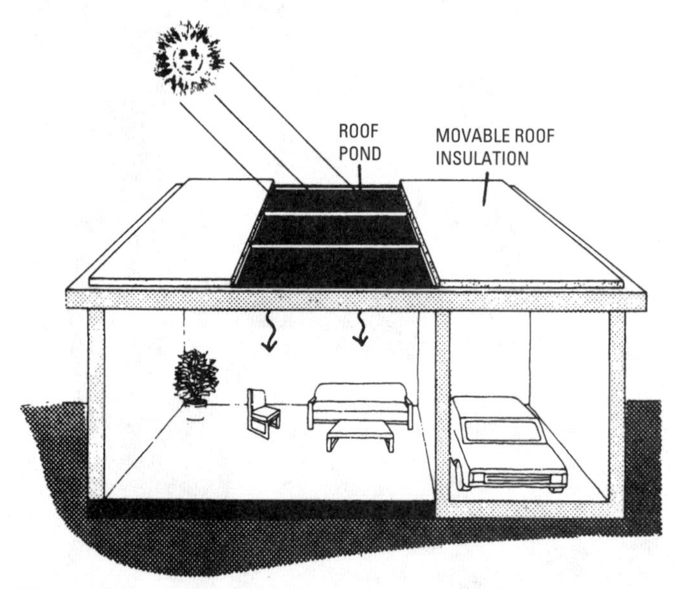

Figure 9-2 A roof pond.

This type of system does have some cooling advantages in the summer. It works well for cooling in parts of the country where significant day-to-night temperature swings take place. The water is cooled down on summer evenings by exposure to the night air. The ceiling water mass then draws unwanted heat from the living and working spaces during the day. This takes advantage of the temperature stratification to provide passive cooling.

This type of system is being tested in California.

Direct Gain

The *direct-gain heating method* uses the sun directly to heat a room or living space (Figure 9-3). The area is open to the sun by using a large windowed space so the sun's rays can penetrate the living space. The areas should be exposed to the south, with the solar exposure working on massive walls and floor areas that can hold the heat. The massive walls and thick floors are necessary to hold the heat. That is why most houses cool off at night even though the sun has heated the room during the day. Insulation must be utilized between the walls and floors and the outside or exterior space. The insulation is needed to prevent the heat loss that occurs at night when the outside cools down.

Woodstoves and fireplaces can be used for auxiliary heat sources with this type of heating system. Systems of this type can be designed with 95 percent passive heating contribution. Overhangs on the south side of the building can be designed to provide shading against unwanted solar gain.

Isolated Gain

In the *isolated-gain* solar heating system, the solar collection and storage are thermally isolated from the living spaces of the building.

The *sunspace* isolated-gain passive building type collects solar radiation in a secondary space. The isolated space is separate from the living space. This design stores heat for later distribution (Figure 9-4).

This type of design has some advantages over the others. It offers separation of the collector-storage system from the living space. It is midway between the direct-gain system (where the living space is the collector of heat) and a mass- or water-trombe system (which collects heat indirectly for the living space).

Part of the design may be an atrium, a sun porch, a greenhouse, or a sunroom. The southern exposure of the house is usually the location of the collector arrangement. A dark tile floor can be used to absorb some of the sun's heat. The northern exposure is protected by a berm and a minimum of window area. The concrete slab is 8 inches thick and will store the heat for use during the night. During the summer, the atrium is shaded by deciduous trees. A fireplace and

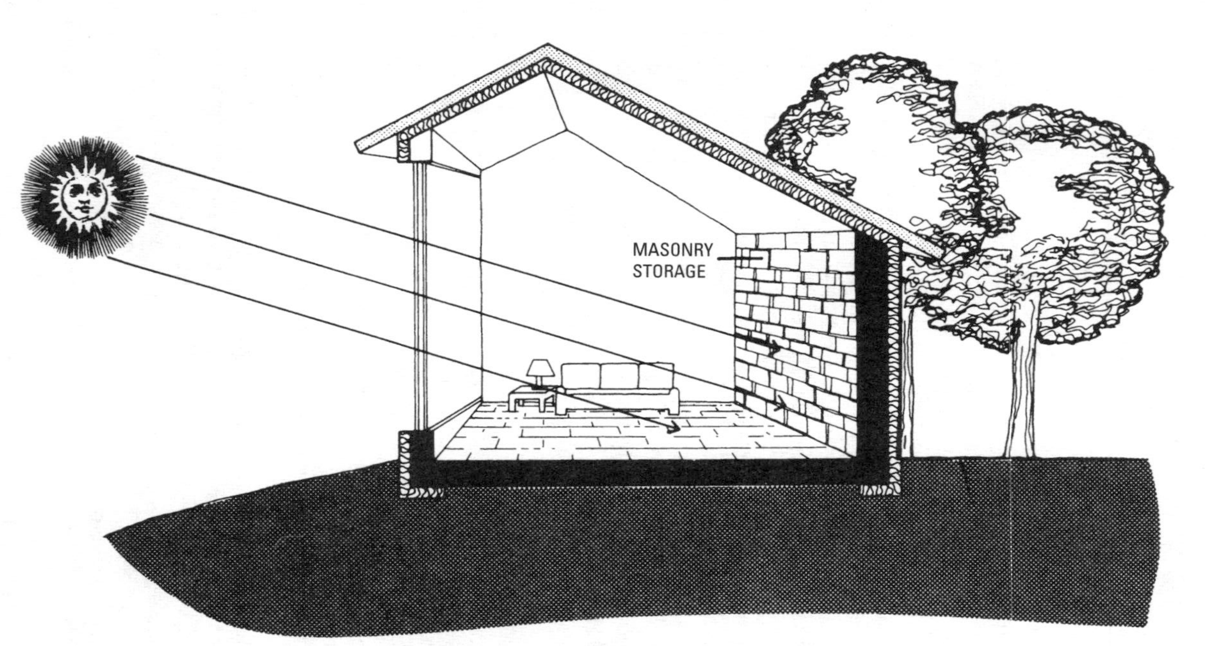

Figure 9-3 Direct-gain passive solar energy system.

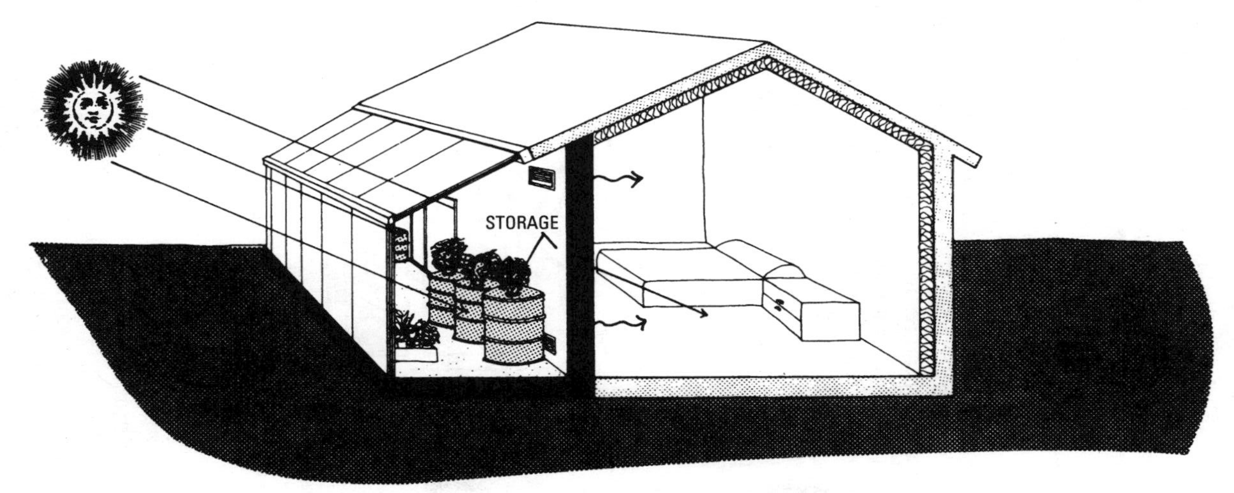

Figure 9-4 This sunspace is isolated from the living space.

small central gas heater are used for auxiliary heat. Performance for the test run has reached the 75- to 90-percent level.

The *thermosiphon* isolated-gain passive building type generates another type of solar-heated building (Figure 9-5). In this type of solar heating system, the collector space is between the direct sunshine and the living space. It is not part of the building. A thermosiphoning heat flow occurs when the cool air or liquid naturally falls to the lowest point (in this case, the collectors). Once heated by the sun, the heated air or liquid rises up into an appropriately placed living space, or it can be moved to a storage mass. This causes somewhat cooler air or liquid to fall again, which in turn causes a continuous circulation to begin. Since the collector space is completely separate from the building, the thermosiphon system resembles an active system. However, the advantage is that there are no external fans or blowers needed to move the heat-transfer medium. The thermosiphon principle has been applied in numerous solar domestic hot water systems. It offers good potential for space-heating applications.

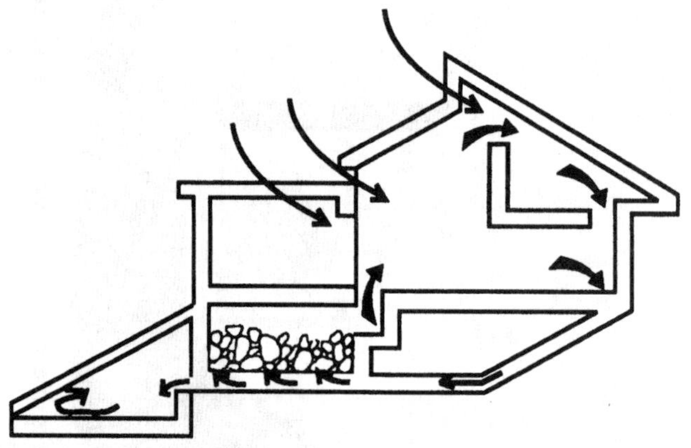

Figure 9-5 Isolated-gain passive heating system: thermosiphon.

Time-Lag Heating

Time-lag heating was used by the American Indians of the Southwest, where it gets extremely hot in the day and cools down during the night. The diurnal (day-to-night) temperature conditions provided a clear opportunity for free or natural heating by delaying and holding daytime heat gain for use in the cool evening hours.

Table 9-1 Heat Capacities of Various Materials

Material	*Specific Heat (BTU/lb/°F)*	*Density (lb/ft³)*	*Heat Capacity (BTU/ft³/°F)*
Water	1.00	62.5	62.5
Concrete	0.27	140	38
Brick	0.20	140	28
Gypsum	0.26	78	20.3
Marble	0.21	180	38
Asphalt	0.22	132	29.0
Sand	0.191	94.6	18.1
White Pine	0.67	27	18.1
Air (75°F)	0.24	0.075	0.018

In those parts of the country where there are significant (20°F to 35°F) day-to-night temperature swings, a building with thermal mass can allow the home itself to delay and store external daytime heating in its walls. The captured heat is then radiated to the building interior during the cool night. Internal heat gains come from people, lights, and appliances. This heat can also be absorbed and stored in the building structure.

Massive or heavy construction of walls, floors, and ceilings are used for this type of solar heating. Because of density, concrete, stone, or adobe has the capacity to hold heat (Table 9-1). As the outside temperature rises during the day, so does the temperature of the building surface. The entire wall section heats up. It will gradually release the stored heat to the room by radiation and convection. Two controls can make time-lag heating systems most effective for passive heating. Two, 4, 8, even 12 hours of delay can be guaranteed by building walls of the right thickness and density. Choosing the right material and thickness can allow you to control what hour in the evening you begin heating. Exterior or sheathing insulation can prevent the heat storage wall from losing its carefully gained heat to the outside, offering more passive heat to the inside. For additional heat, winter sunshine can also be collected and stored in massive walls, provided adequate shading is given to these walls to prevent overheating in the summer.

Uninsulated, massive walls can cause problems with the auxiliary heating system if improperly used. In climates where there is no day-to-night temperature swing, uninsulated massive walls can cause problems. Continually cold or continually hot temperatures outside will build up in heavy exterior walls and will draw heat from the house for hours until these walls have been completely heated from the inside.

This type of house has been built and tested in Denver, Colorado, with a 65-percent passive heating contribution and 60-percent passive cooling contribution.

Underground Heating

The average temperature underground (below the frost line) remains stable at approximately 56°F. This can be used to provide effective natural heating as outside temperatures drop below freezing. The massiveness of the Earth itself takes a long time to heat up and cool down. Its average annual temperature ranges between 55°F and 65°F, with only slight increases at the end of summer and slight decreases at the end of winter. In climates with severe winter conditions or severe summer temperatures, underground construction provides considerably improved outside design temperatures. It also reduces wind exposure. The underground building method removes most of the heating load for maximum energy conservation (Figure 9-6).

Figure 9-6 Underground home.

One of the greatest disadvantages of the underground home is the humidity. If you live in a very humid region of the country, excessive humidity, moisture, and mildew can present problems. This is usually not a problem during the winter season, but can become serious in the summer. Underground buildings cannot take maximum advantage of comfortable outside temperatures. Instead, they are continuously exposed to 56°F ground temperatures. Therefore, if you live in a comfortable climate, there is no need to build underground. You could, however, provide spring or fall living spaces outside the underground dwelling, and use it only for summer cooling or winter heat conservation. Be sure you do not build on clay that swells and slides. In addition, do not dig deep into slopes without shoring against erosion.

An example is a test building that was constructed in Minneapolis. The building was designed to be energy efficient. It was also designed to use the passive cooling effects of the Earth. Net energy savings over a conventional building are expected to be 80 to 100 percent during the heating period and approximately 45 percent during the cooling period.

Passive Cooling Systems

There are seven passive cooling systems. These include natural and induced ventilation systems, desiccant systems, and evaporative cooling systems, as well as the passive cooling that can be provided by night-sky temperature conditions, diurnal (day-to-night) temperature conditions, and underground temperature conditions.

Natural Ventilation

Natural ventilation is used in climates where there are significant summer winds and sufficient humidity (more than 20 percent), so that the air movement will not cause dehydration.

Induced Ventilation

Induced ventilation is used in climate regions that are sunny, but experience little summer wind activity.

Desiccant Cooling

Another name for desiccant cooling is *dehumidification*. This type of cooling is used in climates where high humidity is the major cause of discomfort. Humidity greater than 70 or 80 percent RH (relative humidity) will prevent evaporative cooling. Thus, methods of drying out the air can provide effective summer cooling.

This type of cooling is accomplished by using two desiccant salt plates for absorbing water vapor and solar energy for drying out the salts. The two desiccant salt plates are placed alternately in the living space (where they absorb water vapor from the air) and in the sun (to evaporate this water vapor and return to solid form). The salt plates may be dried either on the roof or at the southern wall, or alternately at east and west walls responding to morning and afternoon sun positions. Mechanized wheels transporting wet salt plates to the outside and dry salt plates to the inside could also be used.

Evaporative Cooling

This type of cooling is used where there is low humidity. The addition of moisture by using pools, fountains, and plants will begin an evaporation process that increases the humidity but lowers the temperature of the air for cooling relief.

Spraying the roof with water can also cause a reduction in the ceiling temperature and cause air movement to the cooler surface. Keep in mind that evaporative cooling is effective only in drier climates. Water must be available for make-up of the evaporated moisture. The atrium and the mechanical coolers should be kept out of the sun, because it is the air's heat you are trying to use. The evaporation process is used, not solar heat. For the total system design, the pools of water, vegetation, and fountain court should be combined with the prevailing summer winds for efficient distribution of cool, humidified air.

Night-Sky Radiation Cooling

Night-sky radiation is dependent on clear nights in the summer. It involves the cooling of a massive body of water or masonry by exposure to a cool night sky. This type of cooling is most effective when there is a large day-to-night temperature swing. A clear night sky in any climate will act as a large heat sink to draw away the daytime heat that has accumulated in the building mass. A well-sized and exposed body of water or masonry, once cooled by radiation to the night sky, can be designed to act as cold storage, draining heat away from the living space through the summer day and providing natural summer cooling.

The roof pond is one natural conditioning system that offers the potential for both passive heating and passive cooling. The requirements for this system involve the use of a contained body of water or masonry on the roof. This should be protected when necessary by moving insulation or by moving the water. The house has conventional ceiling heights for effective radiated heating and cooling. During the summer when it is too hot for comfort, the insulating panels are rolled away at night. This exposes the water mass to the clear night sky. The clear sky absorbs all the daytime heat from the water mass. It leaves the chilled water behind. During the day, the insulated panels are closed to protect the roof mass from the heat. The chilled storage mass below absorbs heat from the living spaces to provide natural cooling for most or all of the day.

This type of cooling offers up to 100 percent passive, nonmechanical air conditioning.

Time-Lag Cooling

This type of cooling has already been described in the earlier section titled, "Time-Lag Heating." It is used primarily in the climates that have a large day-to-night temperature swing. The well-insulated walls and floor will maintain the night temperature well into the day, transmitting little of the outside heat into the house. If 18-inch eaves over all the windows are used, they can exclude most of the summer radiation, thereby controlling the direct heat gain.

Underground Cooling

Underground cooling takes advantage of the stable 56°F temperature conditions of the Earth below the frost line. The only control needed is the addition of perimeter insulation to keep the house temperatures above 56°F. In climates of severe summer temperatures and moderate-to-low humidity levels, underground construction provides stable and cool outside design temperatures, as well as reduced sun exposure to remove most of the cooling load for maximum energy conservation.

Active Solar Heating Systems

Active solar systems are modified systems that use fans, blowers, and pumps to control the heating process and the distribution of the heat once it is collected.

Active systems currently use the units shown in Table 9-2 to collect, control, and distribute solar heat.

Table 9-2 Solar System Units

Unit	*Function*
Solar Collector	Intercepts solar radiation and converts it to heat for transfer to a thermal storage unit or to the heating load.
Thermal Storage Unit	Can be either an air or liquid unit. If more heat than needed is collected, it is stored in this unit for later use. Can be liquid, rock, or a phase-change unit.
Auxiliary Heat Source	Used as a backup unit when there is not enough solar heat to do the job.
Heat Distribution System	Depending on the systems selected, this could be the same as those used for cooling or auxiliary heating.
Cooling Distribution System	Usually a blower and duct distribution capable of using air or liquid directly from either the solar collector or the thermal storage unit.

Operation of Solar Heating Systems

It would take a book in itself to examine all the possibilities and maybe several volumes more to present details of what has been done to date. Therefore, it is best to look at a system that is commercially available from a reputable firm that has been making heating and cooling systems for years (Figure 9-7).

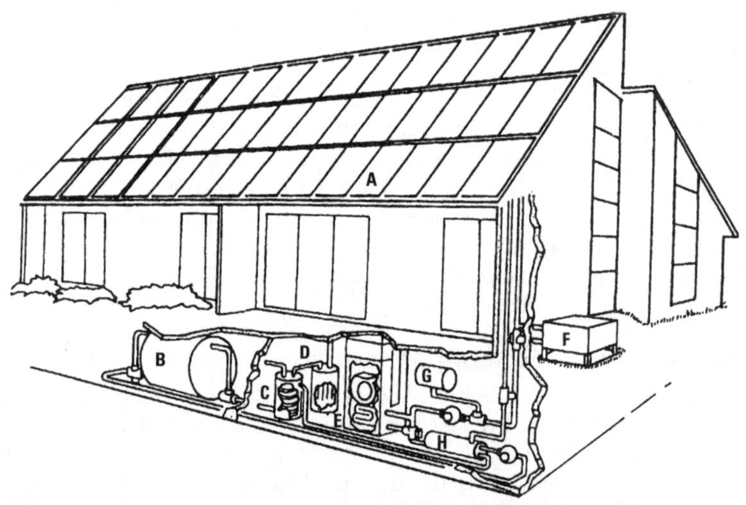

Figure 9-7 Key components in a liquid air solar system: (A) solar collectors; (B) storage tank; (C) hot water heat exchanger; (D) hot water holding tank; (E) space heating coil; (F) purge coil (releases excess solar heat); (G) expansion tank; (H) heat exchanger.

Domestic Water Heating System

The domestic water heating system uses water heated with solar energy. It is more economically viable than whole-house space heating because hot water is required all year around. The opportunity to obtain a return on the initial investment in the system every day of the year is a distinct economic advantage. Only moderate collector temperatures are required to cause the system to function effectively. Thus, domestic water can be heated during less than ideal weather conditions.

Indirect Heating/Circulating Systems

Indirect heating systems circulate antifreeze solution or special heat transfer fluid through the collectors. This is primarily to overcome the problem of draining liquid collectors during periods of subfreezing weather (Figure 9-8). Air collectors can also be used. As a result, there is no danger of freezing and no need to drain the system.

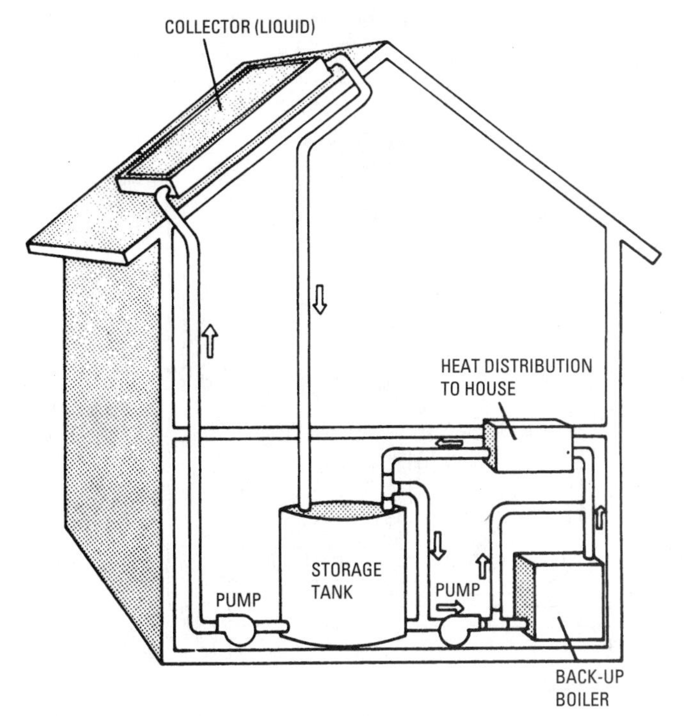

Figure 9-8 Forced-liquid transfer system.

Circulating a solution of ethylene glycol and water through the collector and a heat exchanger is one means of eliminating the problem of freezing. Note that this system requires a heat exchanger and an additional pump. The heat exchanger permits the heat in the liquid circulating through the collector to be transferred to the water in the storage tank. The extra pump is needed to circulate water from the storage tank through the heat exchanger.

The heat exchanger acts as an interface between the toxic collector fluid and the potable (drinkable) water. The heat exchanger must be double-walled to prevent contamination of the drinking water if there is a leak in the heat exchanger.

Air-heating collectors can be used to collect and distribute heat. The operation of this type of system is similar to that of the indirect liquid circulation system (Figure 9-9). The basic difference is that a blower or fan is used to circulate the air through the collector and heat exchanger rather than a pump to circulate a liquid.

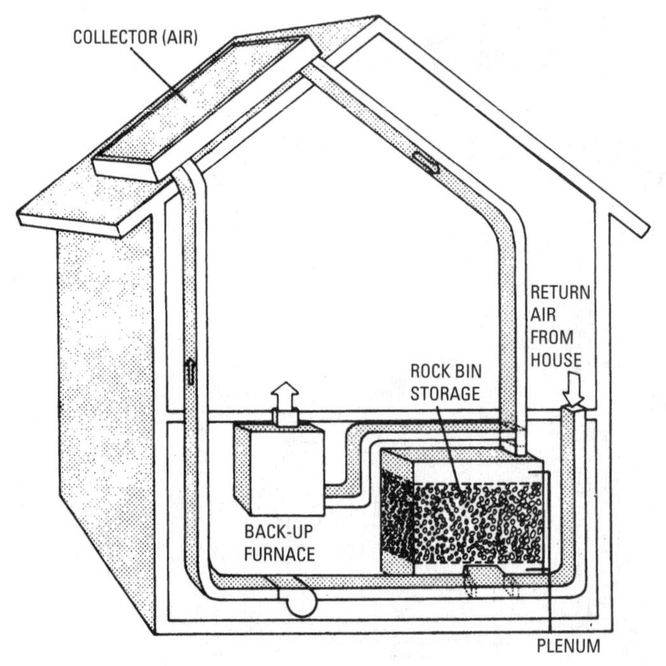

Figure 9-9 Forced-air transfer system.

The air transfer method has the following advantages:

- It does not have any damage caused by a liquid leakage in the collector loop.
- It does not have to be concerned with boiling fluid or freezing in the winter.
- It does not run the risk of losing the expensive fluid in the system.

It does have some disadvantages over the liquid type of system, including the following:

- It requires larger piping between the collector and heat exchanger.

- It requires more energy to operate the circulating fan than it does for the water pump.
- It needs a slightly larger collector.

Summary

There are two ways to classify solar heating systems. The *passive* type is the simplest and relies entirely on the movement of a liquid or air by means of the sun's energy. The *active* system has moving parts to add to the circulation of the heat by way of pumps to push hot water around the system, or fans to blow the heated air and cause it to circulate.

Three concepts are used in the passive heating systems: indirect, direct, and isolated gain. Each of these concepts involves the relationship between the sun, storage mass, and living space. In indirect gain, a storage mass is used to collect and store heat. The storage mass intercedes between the sun and the living space. The three types of indirect-gain solar buildings are mass trombe, water trombe, and roof pond. The direct-gain heating method uses the sun directly to heat a room or living space. The area is open to the sun by using a large windowed space so the sun's rays can penetrate the living space. In the isolated-gain solar heating system, the solar collection and storage are thermally isolated from the living spaces of the building.

The seven passive cooling systems include natural ventilation, induced ventilation, desiccant cooling, evaporative cooling, night-sky radiation cooling, time-lag cooling, and underground cooling.

Active solar systems are modified systems that use fans, blowers, and pumps to control the heating process and the distribution of the heat once it is collected. Active systems currently use units such as solar collectors, thermal storage units, auxiliary heat sources, heat distribution systems, and cooling distribution systems to collect, control, and distribute solar heat.

Review Questions

1. What is the difference between the active and passive solar systems?
2. What is direct and indirect heat gain?
3. Explain one hybrid system.
4. What does the word *trombe* mean?
5. What is isolated gain?

6. What causes the time lag in this type of heating?
7. What is the average temperature underground?
8. What is the greatest disadvantage of underground housing?
9. What is desiccant cooling?
10. What is dehumidification? How is it accomplished?

Chapter 10

Termite Protection

Each year, termites do millions of dollars worth of damage, more than the combined loss from arson, tornadoes, and lightning, and more than twice the damage they caused only 10 years ago. Measures to ensure their control are becoming increasingly necessary. Some of the methods used have not proved to be entirely satisfactory. Certain woods are susceptible to *Lyctus* (powder-post) beetles. If a powdery substance is noticed coming from small holes, these beetles have attacked the wood.

In the case of termite infestation, it is probably best to employ a competent pest-control firm. Termites are tricky. The reproductive pair (king and queen) have been known to live for more than 70 years. The workers and soldiers are blind and live a short life, but they do have a purpose in the ecosystem. Their stomachs have the ability to digest cellulose and put it back into the environment. They help accelerate the decomposition of wood. The queen can get as long as 4 inches when fertile. However, the workers and soldiers are usually small and about the size of ants, although they are not related to ants. There are more than 1700 species of termites. They have adapted to almost every place on Earth with some colonies making external mounds as high as 30 feet in some parts of Australia.

At certain times of the year, termites develop wings, emerge from the ground in swarms like bees, and fly away to form new colonies. Winged termites and winged ants look somewhat alike, but they have definite differences, as shown in Figure 10-1. Termites have a straight body and four milky-white wings, all of equal length and twice as long as the body. Swarming ants also have four wings, but

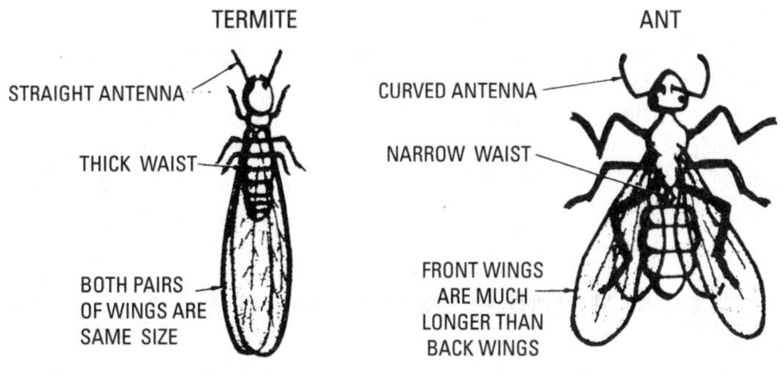

Figure 10-1 Termites and winged ants differ in several respects.

they are transparent and of unequal length, the longest only half again as long as the body. Termites have short, straight, beaded antennae, while the antennae of ants are elbowed, with a bead-like end.

Termites prefer to enter a building through cracks in the foundation or through the cavities in a hollow wall. When these avenues are not available, they will build mud tunnels up from the ground to the wooden parts of the building. These tunnels will be on the surface of the walls (usually on the dark interior side), or behind a pile of lumber, brick, or other materials stacked against the foundation. They are great engineers, however, and have often been known to build hollow tubes up from the ground for a foot or more to reach edible wood. Porch floors consisting of a concrete slab over a fill are fruitful sources of termite troubles, which are exceedingly hard to eradicate. It may necessitate drilling through the floor and injecting heavy doses of toxic chemicals. It may be necessary to do the same where concrete drives or walks contact foundations.

Termites and Their Identification

There are several kinds of termites, but the ones that do the most damage are of the subterranean species. This type lives in the ground, but travels into wood that is in contact with the soil, or into wood that is close enough to the soil to permit building mud tunnels through which they can reach the wood. The presence of these tunnels is sometimes the first sign of a termite infestation. Only a few species of timber are immune to termite damage. Heart cypress and redwood have some slight immunity, but even these woods are occasionally attacked. These termites do not seem to bother very resinous heartwood of long-leaf Southern pine, but this kind of timber is not always available. All species of timbers are immune to termite attack if treated with certain pressurized preservatives.

Termites are often found in stumps or posts and other wood in contact with the soil, but even though such infestations are very near a house, it does not imply that the house is (or will be) infested. Termites seldom leave a location of their own accord, and they work very slowly. There is no need to panic when they are discovered because the damage is already done. A delay of even several months is usually of little consequence.

Construction Methods to Prevent Termite Damage

Termites must maintain contact with the ground to obtain the moisture necessary for their existence. Hence, the first consideration should be to build in a manner that will prevent the entry of termites from the ground:

- The foundations of buildings should be constructed either of masonry or of approved pressure-treated or naturally termite-resistant lumber.
- Where a basement is provided, the foundation walls should be of masonry.
- If unit-block construction is employed (such as brick, tile, or cement blocks), all joints should be well-filled with mortar and the wall topped with a 4-inch cap of concrete. This should be reinforced to prevent cracking when over open-type units.
- The ground within the basement should be sealed over with concrete. Posts should not extend through the floor into the soil, but should rest on concrete footings that extend at least 2 inches above the floor.
- If foundations are built over an earth fill or naturally loose earth, subsequent settlement may cause the joints between the concrete basement floor and foundation walls to open up. Such joints are a probable source of termite entry and should be guarded against by installing mastic or metal expansion joints between the walls and floor.
- Concrete for basement floors and walls should be a dense mixture, and the walls should be reinforced with steel rods at the corners and intersections to tie them together. Improper construction of the basement floor and walls may cause cracks to develop through which termites may gain free access from the earth.
- Windowsills and frames in the basement should not come into direct contact with the ground, nor should leaves or debris be allowed to collect and remain in contact with them.
- Buildings that have no basements should have the sills set a minimum of 18 inches and preferably 24 inches above the excavated ground or natural grade at all points to afford the necessary clearance and good ventilation (Figure 10-2).
- On the exterior, the building clearance to woodwork may be reduced to 8 inches above the finished grade line, provided the foundation walls permit access to occasional inspection for shelter tubes by the homeowner.

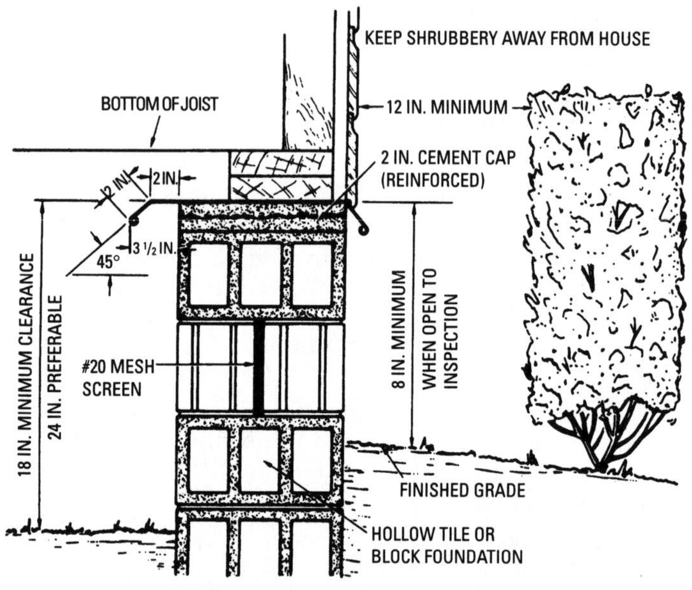

Figure 10-2 Metal shields used in various types of construction to protect against an invasion of termites.

- In the case of solid foundations, ventilation should be provided by allowing not less than 2 square feet of net open area for every 25 linear feet of wall.
- Openings should be screened with 20-mesh noncorroding screening. All lumber that contacts the ground should be pressure-treated with a preservative.

Termite Shields

As pointed out, termites require constant access to soil moisture, and, unless they stay in a moist atmosphere, they soon die. Because of this need for moisture, termites construct shelter tubes of earth or waste material to carry moisture and to act as passageways between the ground and their food supply when it is not in contact with the soil. Destroy or prevent this ground contact, and the termites cannot damage the building.

Termites may build these shelter tubes over the face of stone, concrete, brick, or timber foundations, and along water pipes or similar structures. Such contacts may be prevented by means of a metal shield barrier. This termite shield consists of aluminum firmly inserted and pointed into a masonry joint or under the sill. It projects horizontally at least 2 inches beyond the face of the wall and then is turned downward an additional 2 inches at an angle of 45°. All joints should be locked (and preferably soldered also), with the corners made tight and the outer edge rolled or crimped to give stiffness against bending, as well as to eliminate a sharp edge. The termite shield can also be bedded in asphaltum.

The termite shield should be used on each face of all foundation walls, except that it may be omitted from a face (either interior or exterior) exposed and open to easy and ready inspection. However, around houses and places where shrubbery may partly conceal the wall, and inspection is likely to be infrequent, a modified shield may be used. Here, the horizontal projection is omitted and a 2-inch projection bent downward at 45° is employed. Metal termite shields are required by many codes and by some city ordinances. Figure 10-3 shows some typical termite shields. Figure 10-4 shows a method of inserting shields to prevent the passage of termites by the way of wooden porch steps.

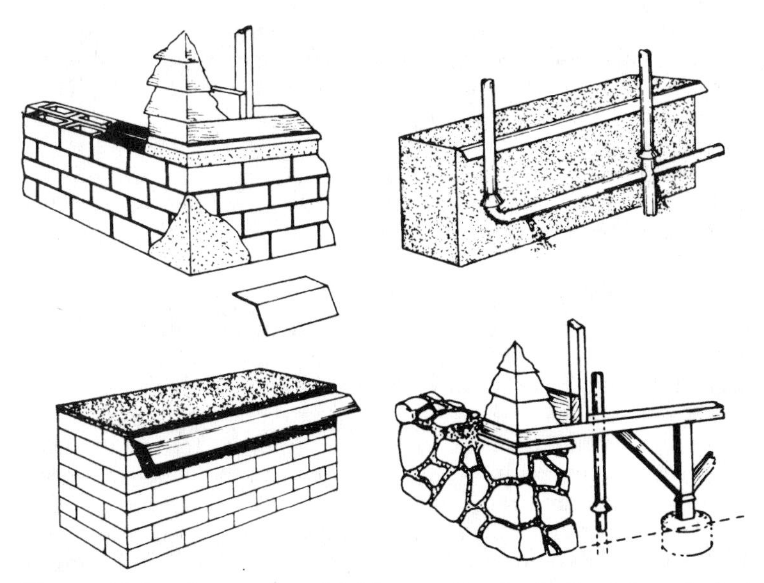

Figure 10-3 A termite shield on top of a hollow foundation wall in a building with no basement.

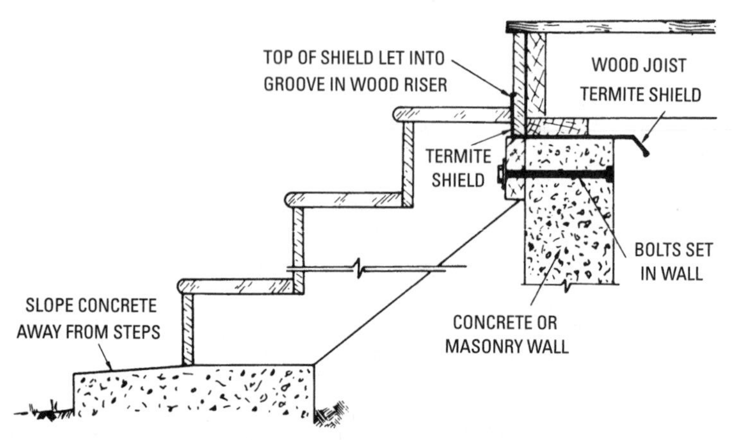

Figure 10-4 Location of a termite shield between wooden steps and a porch.

Detecting and Locating Termite Damage

Except at swarming time, termites are hidden unless their galleries or passageways are broken open. The earth-colored shelter tubes offer a ready means of recognizing their activities. Their work can usually be detected in wood by striking it with a hammer. Solid wood rings clear, while timber or woodwork eaten out by termites will give a dull thud when struck. Striking with an ice pick also determines weakened wood.

Clean-cut holes in books, papers, and clothing are good indications of the presence of termites. Springy basement floors or the softening or weakening of woodwork suggests termite or other damage.

Stopping Termite Damage

The most lasting and effective remedy for termite damage is to replace any wood in or near the basement of the building with concrete. Second in order of effectiveness and durability is to replace wood sections with treated wood or timbers and, in regions of excessive termite damage, to employ protective termite shields. By this means, contact between the colony and the building is permanently broken and relief from termite damage is ensured. This means that joists embedded in concrete, wooden basement floors, and baseboards should be replaced with any type of plain or ornamental concrete. In basement rooms so constructed, movable furniture of wood, and built-in furniture (particularly if resting on concrete footings) can be employed with safety.

To cap and face basement walls of frame buildings, it is rarely necessary to jack up the building. It is usually possible to remove the upper tier of brick or upper portion of the masonry unit in sections and replace with Portland-cement mortar and a suitable cap of slate or mortar. Where poor grades of mortar have been used in masonry walls below the ground, it may be necessary to coat the outside and, if necessary, the inside of the wall with concrete to keep termites from boring through.

Use of Soil Poisons

The use of chemicals is one of the most effective means of termite control. Its success lies in preparing a complete chemical barrier, which leaves no point of entrance for the termites. Two of the most effective and least expensive of the preparations used for this purpose are *chlordane* and *dieldrin*. Both are toxic and have been outlawed in some states. It is best to check local laws before using any chemical.

Summary

There are several types of termites, but the ones that do the most damage are of the subterranean species. This type lives in the ground and travels to wood through tunnels. The presence of mud tunnels is sometimes the first sign of termite infestation. Heart cypress and redwood have a slight immunity, but even these woods are occasionally attacked.

Termites must maintain contact with the ground in order to survive. Moisture must be accessible. The first consideration should be to build in a manner that will prevent the entry of termites from the ground. A foundation of any building should be constructed either of masonry or of approved pressure-treated or naturally termite-resistant lumber.

Where a basement is provided, the foundation walls should be of masonry. If construction of basement walls is of block, all joints should be filled with mortar and the wall topped with a 4-inch cap of concrete.

A termite shield may be installed under the foundation sill to prevent termites from entering the wood. This is a non-corroding metal shield that projects horizontally at least 2 inches beyond the foundation wall and then is turned down at a 45° angle an additional 2 inches. Corners must be tight with the other edge crimped to give stiffness.

The use of soil poisons is another effective means of termite control. Two chemicals most commonly used are chlordane and dieldrin, which are poured into V-shaped trenches against the foundation walls. However, these are illegal in some states, and before using any chemical, the legality of its use should be checked out.

Review Questions

1. What are three ways to prevent termite damage.
2. What are two chemicals used in the ground to prevent termites from entering?
3. What should be done to the top row of blocks in a foundation to prevent termite damage?
4. Name two woods that have a slight immunity to termites.
5. What connection must a termite have with respect to ground?
6. How do termites enter a house?
7. How do termites differ from ants?
8. Which type of termite does the most damage?
9. Why must termites maintain contact with the ground?
10. What is the most lasting and effective remedy for termite damage?

Chapter 11

Maintenance (Preventive and Corrective)

The main principle behind effective maintenance is to control what happens to a building rather than to react to its deterioration. The trick is to notice the building's subtle conditions and then to take appropriate action. Consider the following two examples of how to deal with a screen door.

When a door is broken and doesn't work, it's time to repair it. For example, a screen door falls off its hinges. One of the screens is ripped out and the frame is cracked. It looks like a major repair job. Now that it's broken, you repair it and then reinstall it. This is a reaction to the broken condition of the door, but if a door is out of adjustment and still works, it's an even better time to fix it.

You notice the screen door rubs a little when it closes. Later the door binds and, at times, it will not close. The condition is getting worse, so you look for the cause and tighten a few loose hinge screws. In this case, you correct its condition, a minor task.

Corrective Maintenance

All too often buildings are neglected, so there is a backlog of deferred maintenance to be done. *Corrective maintenance* is recovery from this deferred maintenance. Its purpose is to bring the condition of the building up to an acceptable level.

A corrective maintenance project can be considered "one-time only," as long as the elements involved subsequently receive regular upkeep. Repairing the broken screen door is corrective maintenance.

Preventive Maintenance

Preventive maintenance is action to improve conditions before they become unacceptable. It is a continuous process. Activities such as housekeeping are repeated at daily or weekly intervals, and the pattern is easy to see. However, reglazing windows might only be done every 20 or 40 years, so a pattern is more difficult to recognize and control.

An aim of preventive maintenance is to slow down deterioration as much as possible. This is really what maintenance is all about. Exterior paint continues to weather away no matter what we do. The loss of the paint is acceptable because it prevents deterioration of the clapboards beneath. To preserve the clapboards you renew the paint every 5 to 15 years. Tightening the screen door screws is another example of preventive maintenance.

Checklist

A house that is well-constructed, with adequate attention to construction details and to the choice of materials used to build the house, will need far less maintenance than one that is not well-built. It is indeed discouraging to the house owner to begin repair and maintenance almost before moving in! An extra $40 used for rust-resistant nails on the siding, for example, may save $500 by requiring less frequent painting.

The following sections describe conditions and areas to check for problems, and how to solve them.

Basement

The basement may sometimes be damp for several months after the house has been built. In most cases, this moisture comes from masonry walls and floors. It will progressively disappear. In cases of persistent dampness, however, the owner should check various areas to eliminate any possibilities for water entry.

The following areas may be the source for some of the trouble caused by the entry of moisture:

- Check the drainage at the downspouts. The final grade around the house should be away from the building, and a splash block should be provided to drain water away from the foundation wall.
- Some settling of the soil may occur at the foundation wall and form water pockets. These areas should be filled and tamped so that surface water can drain away.
- Some leaking may occur in a poured concrete wall at the tie wires. These usually seal themselves, but larger holes should be filled with a cement mortar. Clean and slightly dampen the area first for good adhesion.
- Concrete block or other masonry walls exposed above grade often show dampness on the interior after a prolonged rainy spell. There are concrete paints on the market that increase the resistance to moisture seepage. Waterproofing materials may also be used on the exterior walls.
- For moisture absorption there should be at least a 6-inch clearance between the bottom of the siding and the finish grade to prevent moisture absorption. Shrubs and foundation plantings should also be kept away from the wall to improve circulation

and drying. In lawn sprinkling, do not allow the water to spray against the walls of the house.

- Check areas between the foundation wall and the sill. Any openings should be filled with caulk. This filling will decrease heat loss and prevent entry of insects into the basement.
- Dampness in the basement in the early summer months is often augmented by opening the windows for ventilation during the day. This will allow warm, moisture-laden outside air to enter. The lower temperature of the basement will cool the incoming air and frequently cause condensation to collect and drip from cold-water pipes and collect on colder parts of the masonry walls and floors. To air out the basement, open the windows during the night.

In aggravated cases, heating the basement to raise the temperature 5° to 10° (with the windows closed) has proved helpful. Heating lowers the humidity, warms the cold masonry, and dries up the moisture. On cool days, no heat is required, and the windows may be opened. As the summer advances and the masonry warms up, condensation will not usually occur on the walls but may occur on the cold-water pipes. Wrapping such pipes has proved helpful in reducing condensation and drip.

A desiccant (such as calcium chloride) is sometimes used to lower the humidity in basements. Dehumidifiers are also available in a variety of sizes. Such units are generally enclosed in a cabinet and are operated by electricity. The doors and windows should be closed at all times during dehumidification. Otherwise, the moisture in the entering air will continually replace the moisture extracted by the desiccant or dehumidifier.

Crawl-Space Area

The crawl-space area should be checked as follows:

- If the house is located in a termite area, be sure to make annual inspections (preferably in the late spring or early summer) for termite activity or damage. These inspections are necessary if the house or a part of the house is built over a crawl space. Termite tubes on the walls or piers are an indication of their activity. A well-constructed house will have a termite shield under the wood sill with a 2-inch extension on the interior. Examine the shield for proper projection, and any cracks in the foundation walls, as these cracks form good channels for termites to enter.

- While in this crawl space, it is a good idea to check the area for any decay that may be occurring in the girders or joists, as well as for signs of condensation on wood members of the floor framing. Use a penknife to test questionable areas for rot and decay. There are products (such as Git Rot) for repairing rotted wood. This is available at marine supply stores.
- If the crawl space seems damp and the ground moist, it may be because of lack of proper ventilation. The crawl-space ventilators should be of adequate size and should be so located that there is crosscirculation. The use of a soil cover (55-pound saturated felt or heavier, or polyethylene sheeting) on the ground of the crawl space will prevent much of the soil moisture from entering the area.

Roof and Attic

The roof and attic should be checked as follows:

- If there are a few humps on asphalt-shingle tabs, they most likely are caused by nails that have not been driven into solid wood. Remove such nails and replace with others driven into sound wood. There may be a line of buckled shingles along the roof. This ordinarily is caused by applying the shingles to roof boards that are not dry and to roof sheathing that varies in thickness. The shrinkage of the wood in the wide boards between the nails will also cause these humps by buckling the shingles. Time and hot weather tend to reduce this condition.
- A dirt streak down the gable end of the house is often caused by rain entering the rake molding from shingles that have insufficient projection. A cant strip would ordinarily have prevented this situation.
- In winters of heavy snows, ice dams may form at the eaves, and these often result in water entering the cornice and walls of the house. The immediate remedy is to remove the snow on the roof for a short distance above the gutters and, if necessary, in the valleys. Additional insulation between heated rooms and roof space, and increased ventilation, will help to decrease the melting of snow on the roof and thus minimize formation of ice at the gutters and in the valleys. Deep snow in valleys also sometimes forms ice dams that cause water to back up under shingles and valley flashing. You can also get heating cables to install in gutters.

- Roof leaks are often caused by improper flashing at the valley, ridge, or around the chimney. Observe these areas during a rainy spell to discover the source. Water may travel many feet from the point of entry before it drips off the roof members.
- The attic ventilators serve two important purposes: that of summer ventilation (as a means of lowering the attic temperature to improve comfort conditions in the rooms below) and that of winter ventilation (to remove moisture that may work through the ceiling and condense in the attic space). The ventilators should be open in both winter and summer.

To check for sufficient ventilating area during cold weather, examine the attic after a prolonged cold period. If nails protruding from the roof are heavily coated with frost, it is evident that ventilation is not sufficient. Frost may also collect on the roof sheathing, first appearing near the eaves on the north side of the roof. Increase the size of the ventilators or place additional ones in the protected underside of the cornice. This will improve air movement and circulation.

Exterior Walls

One of the major problems in maintenance of a wood-covered house is the exterior paint finish. There are a number of reasons for paint failures, many of them known; others are not yet thoroughly investigated. One of the major causes of paint failure is moisture in its various forms. Quality of paint and method of application are other reasons. Chapter 5, "Painting," discusses correct methods of application, types of paint, and the problems encountered.

Another problem with exterior walls that the owner may encounter in a house is if steel nails have been used for the application of the siding, rust spots may occur at the nail head. Such rust spots are quite common where nails are driven flush with the heads exposed. Spotting is somewhat less apparent where steel nails have been set and puttied. In the case of flush nailing, the spotting may be minimized by setting the nail head below the surface and puttying. The puttying should be preceded by a priming coat. Take care of these nails just before the house requires repainting.

Brick and other types of masonry are not always waterproof, and continued rains may result in damp interior walls or wet spots where water has soaked through. If this trouble persists, it may be a good idea to use a waterproof coating over the exposed surfaces. Coatings can be obtained for this purpose. Caulking is usually required where a change in materials occurs (such as that of wood siding abutting against brick chimneys or walls). The wood should always have a prime coating of paint for proper adhesion of the caulking compound. Caulking guns with cartridges can be obtained and are the best means of waterproofing these joints.

Rainwater flowing down over wood siding may work through butt and end joints and sometimes may work up under the butt edge by capillary action. Coating edges with paint helps, and painting under the butt edges at the lap adds mechanical resistance to water intrusion. However, moisture changes in the siding cause some swelling and shrinking that may break the paint film. Treating the siding with a water repellent before it is applied is an effective method of reducing capillary action. For houses already built, the water repellent could be applied under the butt edges of bevel siding or along the joints of drop siding and at all vertical joints. Excess repellent on the face of painted surfaces should be wiped off.

Interior Walls

In a newly constructed house, many small plaster cracks may develop during or after the first heating season. These cracks are usually caused by the drying and shrinking of the structural members. For this reason, it is advisable to wait for a part of the heating season before painting, if the walls are plastered. These cracks can then be filled before painting. Because of the curing period ordinarily required for plastered walls, it is not advisable to apply oil-base paints until at least 60 days after plastering is completed. Latex paints can be applied without waiting.

Large plaster cracks often indicate a structural weakness in the framing. One of the common areas that may need correction is around a basement stairway. Framing may not be adequate for the loads of the walls and ceilings. In such cases, the use of an additional post and pedestal may be required to correct this fault. Inadequate framing around fireplaces and chimney openings, and joists that are not doubled under partitions, are other common sources of weakness.

Moisture on Windows

Points to be noted with respect to moisture on windows are as follows:

- During cold weather, condensation (and, in cold climates, frost) will collect on the inner face of single-glazed windows. Water from the condensation or melting frost runs down the glass and soaks into the wood sash to cause stain, decay, and paint failure. The water may rust steel sash. To prevent such condensation, the window should be provided with a storm sash. Double-glazing will also minimize this condensation.

- Occasionally, in very cold weather, frost may form on the inner surfaces of the storm windows. This may be caused by
  - Loose-fitting window sash. This may allow moisture to enter the space between the window and storm sash.
  - High relative humidity in the living quarters.

Generally, the condensation on a storm sash does not create a maintenance problem, but it may be a nuisance. Weatherstripping the inner sash offers increased resistance to moisture flow and may prevent this condensation. Lower relative humidity in the house is also helpful.

Moisture on Doors

Condensation may collect on single exterior doors during severe cold periods for the same reasons as described for windows. Here again, the water may cause damage to the door and to the finish. Storm doors offer the most practical means of preventing or minimizing such condensation. The addition of storm doors will also decrease the warping of the exterior doors.

Floors

A finish floor that has been improperly laid is a source of trouble. This flooring may have been laid with varying moisture contents in the boards, or at too high a moisture content. Cracks or openings in the floor appear during the heating season as the flooring dries out. If the floor has a few large cracks, one expedient is to fit matching strips of wood between the flooring strips and to glue them in place. In severe cases, it may be necessary to replace sections of the floor or to refloor the entire house. Another method would be to cover the existing flooring with a thin flooring, $^{5}/_{16}$ -or $^{3}/_{8}$ -inch thick. This would require removal of the base shoe, fitting the thin flooring around doorjambs, and perhaps sawing off the door bottoms.

Summary

Preventive maintenance and alertness to problems will often prevent a major repair bill. A damp basement will sometimes occur for several months after a house has been completed, and, in most cases, it will progressively disappear. In cases of persistent dampness, the owner should check various areas to eliminate this problem. Checking drainage downspouts for proper grade to carry away drain water is one of the first items on the preventive maintenance list.

Some settling of the soil may occur at the foundation wall and form a water pocket.

If a house is designed with a crawl-space area, it should be checked for termite activity or damage. A well-constructed house will have a termite shield under the wood sill. This should also be checked for proper projection. Crawl-space ventilators should be checked for proper air circulation, as well as any obstruction, which could prevent circulation.

The roof and attic should be inspected for humps in the asphalt shingles. These are an indication of loose nails. All flashing around chimneys, valleys, and ridges should be checked for possible leaks. Attics should be checked for proper ventilation and leaks in the roof.

Interior and exterior walls should be checked periodically for cracks, paint peeling, and air leaks. Windows should be checked for air leaks and freedom of operation. Floors should be checked for cracks or openings, which appear during the heating season. Floor joists should be checked for cracks or decay.

Review Questions

1. How often should a frame house be painted?
2. Why does plaster crack above windows and door openings?
3. How can dampness be corrected in a crawl space?
4. What should be checked when inspecting a roof?
5. Why does a wood floor crack or buckle after being installed?
6. Why are basements sometimes damp or moist six months after the move-in?
7. Preventive maintenance is ______ to improve conditions before they become unacceptable.
8. If the crawl space seems damp and the ground moist, it may be because of lack of proper ______.
9. Condensation on a storm sash does not create a maintenance problem, but it may be a ______.
10. What causes the occasional frost that forms on the inner surfaces of the storm windows?

Chapter 12

Chimneys and Fireplaces

Chimneys and fireplaces are usually constructed by masons rather than by carpenters. However, many tradespeople are adept at more than one skill. In any case, it is important for the carpenter to understand chimney construction, since special wood framing and detailing are needed to meet critical fire safety codes.

One national standard code is *Chimneys, Fireplaces, Vents, and Solid Fuel Burning Appliances,* developed by the National Fire Protection Association. While many local jurisdictions use this standard as a model, always check with local officials to understand how local code requirements may vary.

Chimneys

The proper construction of a chimney is more complicated than it seems. It must be designed to suit its users and blend with the architectural style of the building. It must be mechanically strong, contain and control burning fires, and channel smoke out of the building safely.

Conditions for Proper Draft

The power that produces a natural draft in a chimney is caused by the small difference in weight between the hot gases in the flue and the colder air outside. The hot gases are lighter and will, therefore, rise. This provides a natural draft up the chimney.

When flow within the flue reverses (bringing smoke down into the building), it is called a *downdraft.* This can be more than a nuisance, since the hot gases also in the flue can be poisonous. Downdrafts happen when wind sweeps across the roof and swirls down on top of the chimney. One cause of the downsweep of the wind is when the peak of the roof extends higher than the top of the chimney. To prevent this condition, ensure that the top of the chimney is at least 2 feet above the nearby roof, as shown in Figure 12-1(A). The height above a flat roof should not be less than 3 feet, as shown in Figure 12-1(B). Large trees or hills near a dwelling can also cause downdrafts. A chimney cap with flue openings at the sides will sometimes cure severe downdraft problems.

Another cause of poor draft is a restriction inside the chimney that offers resistance to the free flow of gases. This restriction could be a design fault such as an abrupt offset (Figure 12-2).

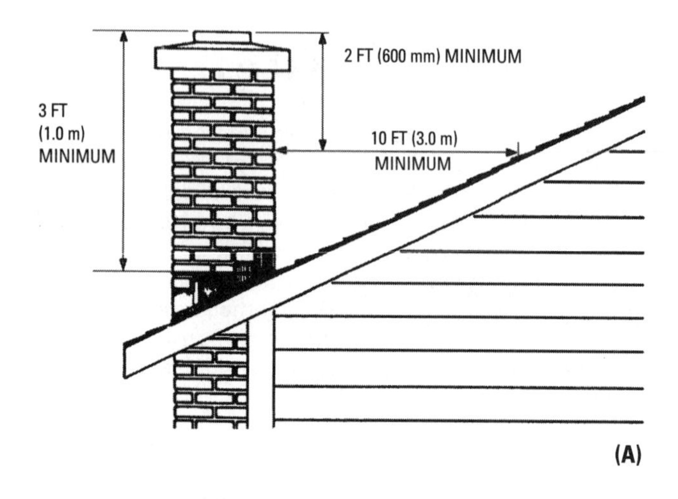

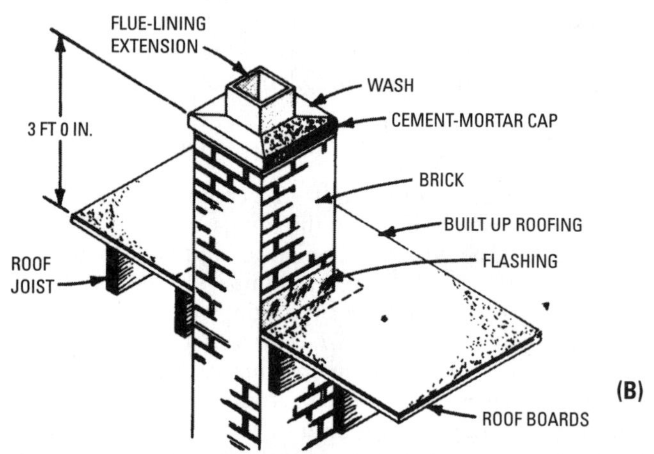

Figure 12-1 Chimney heights: (A) above a gable roof and (B) above a flat roof.

Foundations

When considering the weight of even the smallest chimney, it is clear that a good foundation is necessary to support it properly. The foundation should go as deep as necessary to rest on hardpan or other stable and undisturbed soil. If the chimney is built on an outside wall, the foundation must extend to below the frost line. This may be as deep as 4 feet in some northern sections of the United States, but the depth varies in different localities.

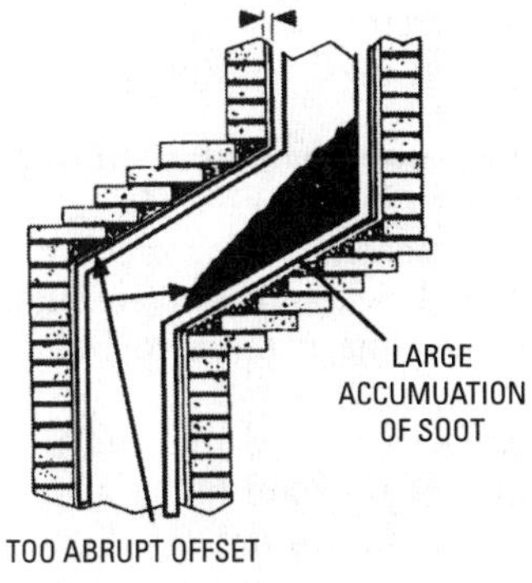

(A) POOR: Extreme offset. Accumulation of soot and ashes may eventually block the chimney or ignite, causing a severe chimney fire. May develop cracks leading to structural failure.

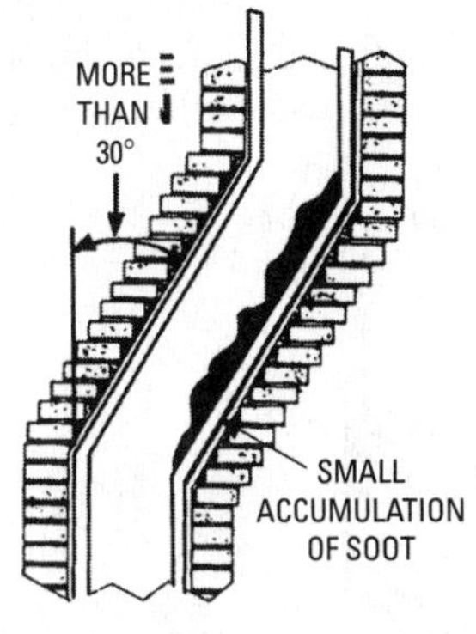

(B) BETTER: Gradual offset. Minor soot accumulation will not block chimney but may still be difficult to clean and maintain.

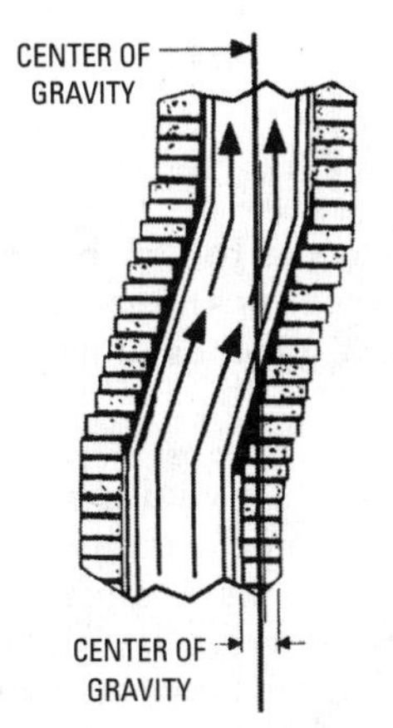

(C) BEST: Limited offset. Centerline of flue above falls within center of flue wall below. Chimneys should be built with as little offset as possible. This not only provides a better draft but is more structurally stable.

Figure 12-2 Offset in a flue must not be too abrupt.

The foundation should extend at least 6 inches on all sides of the chimney and should be of reinforced concrete.

Construction

A straight flue is the most efficient and easiest to build. If an offset is necessary, follow the guidelines shown in Figure 12-2(C).

Common brick has been the standard masonry unit for complex chimney construction. However, concrete brick and concrete block are used extensively today to reduce the time it takes to build a chimney and to reduce materials costs.

Bricks or blocks are laid up in courses around the flue (Figure 12-3). The flue is lined with special heat-resistant materials that should never be omitted. Typically, the lining is a fireclay flue tile, but firebrick or a refractory cement lining could be used. The flue lining runs up the entire height of the chimney and protects the outer courses from corrosive flue gases and intense heat (Table 12-1).

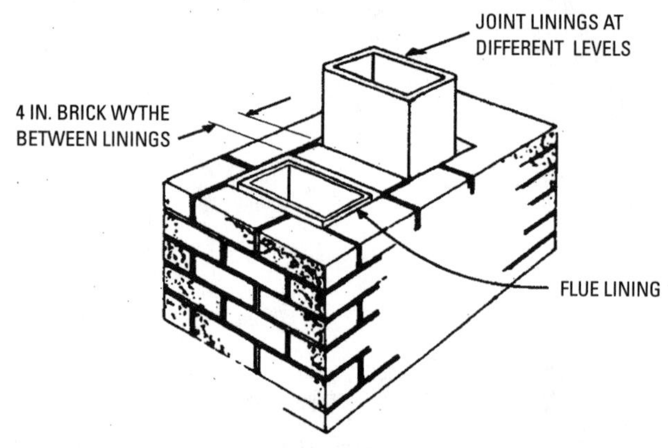

Figure 12-3 Flue lining should extend to the top of the chimney. Stagger lining joints so they don't occur at the same level. Separate two or more flues with at least 4 inches of masonry called a wythe.

Most masonry materials expand and contract with changes in temperature. The flue lining normally expands and contracts differently than the outer course of brickwork, because it is made of a different material and it gets much hotter. With fireclay flue tiles, leave a 1-inch air space between the tile and the outer brickwork. This reduces stress between the tile and the brickwork because each can move somewhat independently.

Table 12-1 Rectangular and Round Flue Linings

Rectangular Flues		***Round Flues***	
Outside Dimensions (Inches)	***Net Inside Area (Square Inches)***	***Inside Diameter (Inches)***	***Net Inside Area (Square Inches)***
$4^1/_2 \times 8^1/_2$	25	6	28.3
$4^1/_2 \times 13$	40	7	38.5
$4^1/_2 \times 18$	57	8	50.3
$8^1/_2 \times 8^1/_2$	52	9	63.6
$8^1/_2 \times 13$	80	10	78.5
$8^1/_2 \times 18$	113	12	113.
13×13	126	15	176.7
13×18	169	18	254.5
18×18	240	20	314.2
		24	452.4

To give wind blowing across the top of the chimney an upward sweep, project the lining about 5 inches above the brickwork, as shown in Figure 12-3 and Figure 12-4. Then form a cap of cement within 2 inches of the top of the lining. The cement cap should not be bonded to the flue so that the flue can rise because of heat expansion without cracking the top of the chimney.

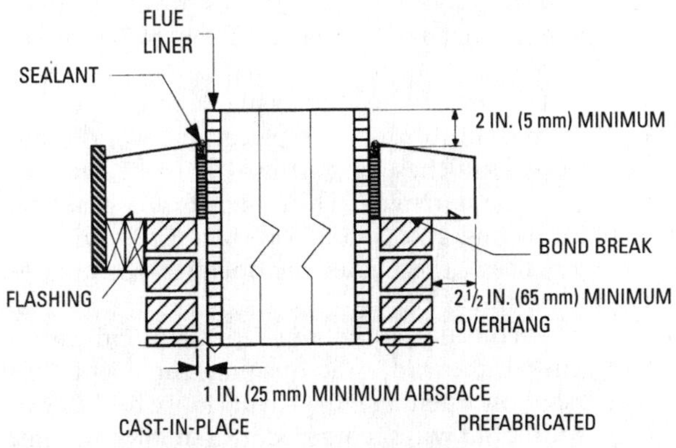

Figure 12-4 Cap details. A prefabricated cap will last longer than a cast-in-place cap. Flexible sealant allows for liner movement and helps the overhang protect the chimney from water penetration.

Fireplaces

Since the energy crunch, fireplaces have become popular. However, they normally do not lower fuel bills, because they are not efficient. If you want a fireplace, you should choose it more for the cheerful and cozy feeling that a crackling blaze in a fireplace gives. Fireplaces are considered appropriate in living rooms, dining rooms, and bedrooms.

There are a number of design features to observe to ensure that the fireplace performs satisfactorily. Fireplace size should match the room in which it is to be used, both from the standpoint of appearance and operation. If the fireplace is too small, it will not adequately heat the room. If it is too large, a fire that fills the combustion chamber will be much too hot for the room and will waste fuel.

The location of the chimney usually determines the location of the fireplace. It is often governed by structural considerations only. A fireplace should be located to provide a reasonable amount of seclusion and should not be near doors that carry traffic from one part of the house to another.

Design

In designing a fireplace, the characteristics of the fuel to be burned should be considered. In addition, the design should harmonize with the room in both proportion and detail. In the days of the early settlers of our land, wood was plentiful and fireplaces with openings as large as 7 feet wide and 5 feet high were common. These were often used for cooking, as well as heating. They consumed huge amounts of fuel and were more likely to be smoky than not.

Where cordwood (4 feet long) is readily available and practical, a 30-inch-wide opening is desirable for a fireplace. The wood is cut in half to provide pieces 24 inches long. If coal is to be burned, however, the opening can be narrower. Thirty inches is a practical height for any fireplace having an opening less than 6 feet wide. It will be found that, in general, the higher the opening, the greater the chance the fireplace will smoke.

Another dimension to consider is the depth of the combustion chamber. Again, in general, the wider the opening, the deeper the chamber should be. A shallow opening will provide more heat than a deep one of the same width, but will accommodate a smaller amount of fuel. Thus, a choice must be made as to which is the more important: a greater depth (which permits the use of larger logs that burn longer) or a shallow depth (which takes smaller pieces of wood but provides more heat).

In small fireplaces, a depth of 12 inches will provide adequate draft if the throat is constructed properly, but a minimum depth of 16 to 18 inches is advisable to lessen the danger of burning fuel falling out on the floor. Screens of suitable design should be placed in front of all fireplaces.

A fireplace 30 to 36 inches wide is usually suitable for a room having 300 square feet of floor area. The width should be increased for larger rooms, with all other dimensions being changed to correspond to the sizes listed in Table 12-2. Following these dimensions will result in a fireplace that operates correctly.

Modified Fireplaces

Metal fireplaces are available in many different sizes. They are set in place and concealed by the brick or stone used to lay up the fireplace. Figure 12-5 shows one of these units. With this type of fireplace, the warm air is circulated throughout the room by the action of the air inlets and outlets provided in the metal unit. Cool air is drawn into the inlets at the bottom of the unit and is heated by contact with the metal sides. This heated air rises and is discharged through the outlets at the top. The inlets and outlets are connected to registers that may be located at the front or ends of the fireplace, or even in another room.

One advantage of this type of modified fireplace is that the firebox is correctly designed and proportioned and has the throat damper, smoke shelf, and chamber already fabricated. All this provides a foolproof form for the masonry. It greatly reduces the risk of failure and practically ensures a smokeless fireplace. It must be remembered, however, that even though the fireplace is built correctly, it will not operate properly if the chimney is inadequate.

The quantity and temperature of the heated air discharged from the registers in this type of fireplace will circulate heat into the coldest corners of a room and will deliver heated air through ducts to adjoining or overhead rooms. For example, heat could be diverted to a bathroom from a fireplace in the living room. The amount of heat delivered by some of the medium-size units is equivalent to that from 40 square feet of cast-iron radiators in a hot water heating system. This is sufficient to heat a 15-foot × 18-foot room to 70°F when the outside temperature is 40°F. Additional heat can be obtained with some models by using circulation fans placed in the ductwork.

Figure 12-6 shows another type of modified fireplace in which the circulated air is drawn in from outside through an air inlet, *A*. The air is then heated as it rises through the back heating chamber,

Table 12-2 Recommended Dimension for Fireplaces

Opening			Minimum Back Wall (Horizontal) (Inches)	Vertical Back Wall (Inches)	Inclined Back Wall (Inches)	Outside Dimensions of Standard Rectangular Flue Lining (Inches)	Inside Diameter of Standard Round Flue Lining (Inches)
Width (Inches)	Height (Inches)	Depth (Inches)					
24	24	16–18	14	14	16	$8^1/_2 \times 8^1/_2$	10
28	24	16–18	14	14	16	$8^1/_2 \times 8^1/_2$	10
24	28	16–18	14	14	20	$8^1/_2 \times 8^1/_2$	10
30	28	16–18	16	14	20	$8^1/_2 \times 13$	10
36	28	16–18	22	14	20	$8^1/_2 \times 13$	12
42	28	16–18	28	14	20	$8^1/_2 \times 18$	12
36	32	18–20	20	14	24	$8^1/_2 \times 18$	12
42	32	18–20	26	14	24	13 × 13	12
48	32	18–20	32	14	24	13 × 13	15
42	36	18–20	26	14	28	13 × 13	15
48	36	18–20	32	14	28	13 × 18	15
54	36	18–20	38	14	28	13 × 18	15
60	36	18–20	44	14	28	13 × 18	15
42	40	20–22	24	17	29	13 × 13	15
48	40	20–22	30	17	29	13 × 18	15
54	40	20–22	36	17	29	13 × 18	15
60	40	20–22	42	17	29	18 × 18	18
66	40	20–22	48	17	29	18 × 18	18
72	40	22–28	51	17	29	18 × 18	18

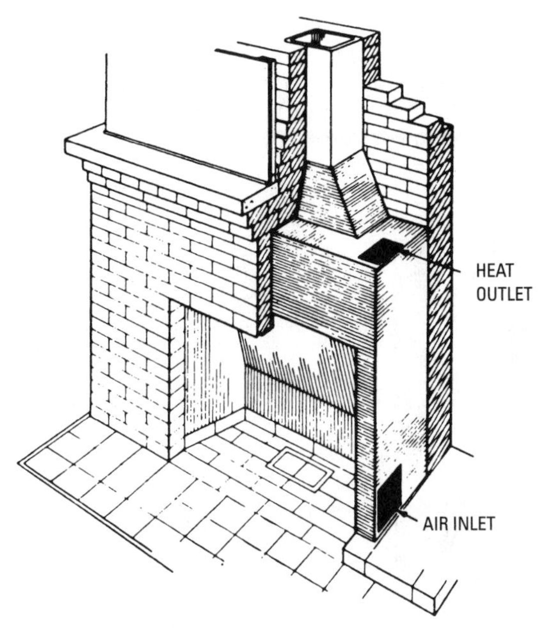

Figure 12-5 A metal fireplace unit practically ensures a smokeless fireplace.

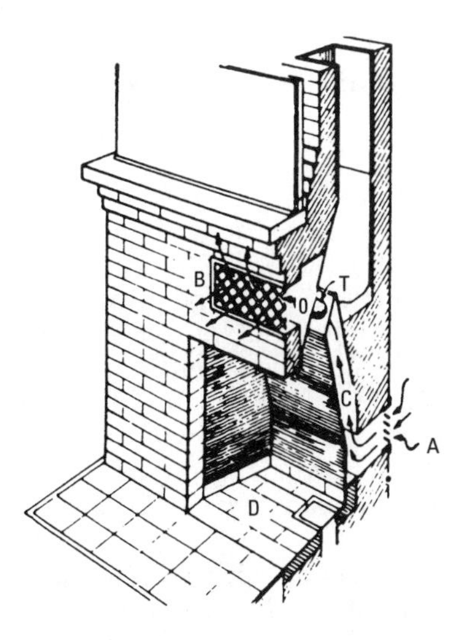

Figure 12-6 A type of modified fireplace in which the air to be heated is drawn in from outside.

C, and tubes, *T*, being discharged through register *B*. Combustion air is drawn into the fireplace opening, *D*, and passes between the tubes and up the flue. Dampers are provided to close the air inlet and regulate the amount of combustion air allowed to pass into the flue. Figure 12-7 shows a cross-section of this type of modified fireplace.

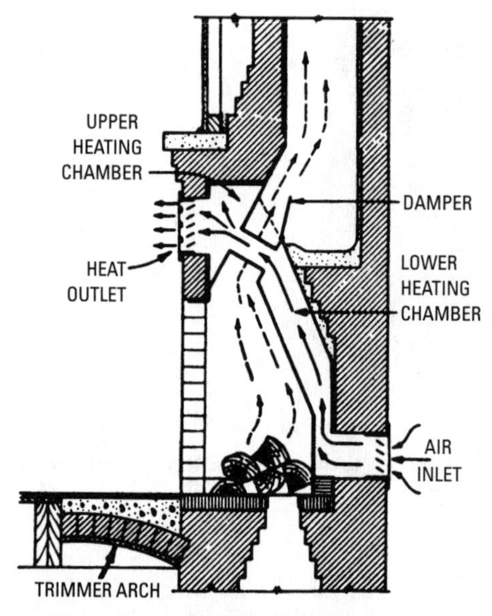

Figure 12-7 A cross-sectional view of the modified fireplace shown in Figure 12-6.

Prefabricated Fireplaces

Prefabricated fireplaces are available in many designs and price ranges. These units fill the need for a low-cost fireplace that eliminates expensive masonry construction. This type of fireplace can be free-standing, mounted flush with the wall, or recessed into it, depending on the style selected. The unit shown in Figure 12-8 has a cantilever hearth 15 inches above floor level and can burn wood up to 27 inches long. The outer shell is steel with stainless steel trim, while the firebox is formed of high-impact ceramic material. Figure 12-9 shows a prefabricated fireplace that hangs from the ceiling.

The complete unit (except for the trim) is prime-coated and ready to be painted with any interior paint to harmonize with the

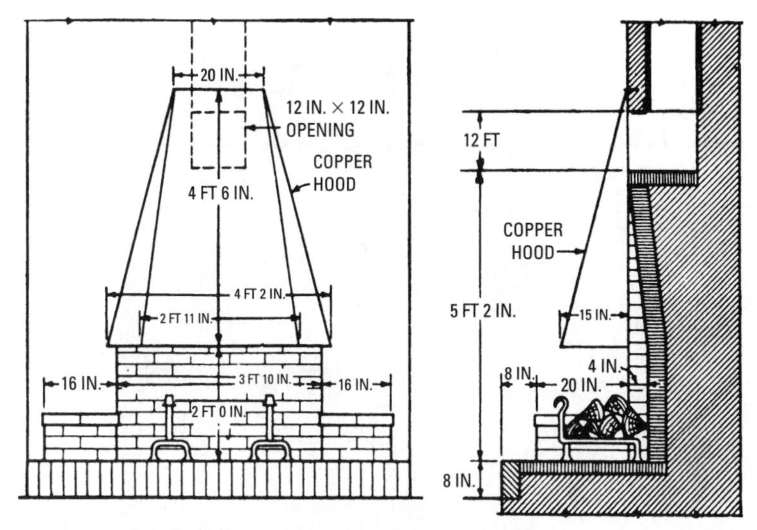

Figure 12-8 A prefabricated fireplace that requires a minimum of masonry.

decorating scheme of the room. A flexible hearth screen is provided to prevent sparks from flying out of the firebox into the room.

Construction

Fireplace construction is an exacting art. Preformed metal units may be used, or the complete fireplace can be built of masonry materials. Figure 12-10 shows the details of a typical fireplace. This, along with the following essentials, should prove helpful in the design and construction of any type of fireplace:

- The flue must have sufficient cross-sectional area.
- The throat must be constructed correctly and have a suitable damper.
- The chimney must be high enough to provide a good draft.
- The shape of the firebox must be such as to direct a maximum amount of radiated heat into the room.
- A properly constructed smoke chamber must be provided.

Dimensions

Table 12-2 lists the recommended dimensions for fireplaces of various widths and heights. If a damper is installed, the width of the

Figure 12-9 A free-standing fireplace can be located almost anywhere.

opening will depend on the width of the damper frame and slope of the back wall.

The width of the throat is determined by the opening of the hinged damper cover. The full damper opening should never be less than the cross-sectional area of the flue. When no damper is used, the throat opening should be 4 inches for a fireplace that does not exceed 4 feet in height.

Hearths

The hearth in conventional fireplaces should be flush with the floor to permit sweepings to be brushed into the fireplace opening. If a

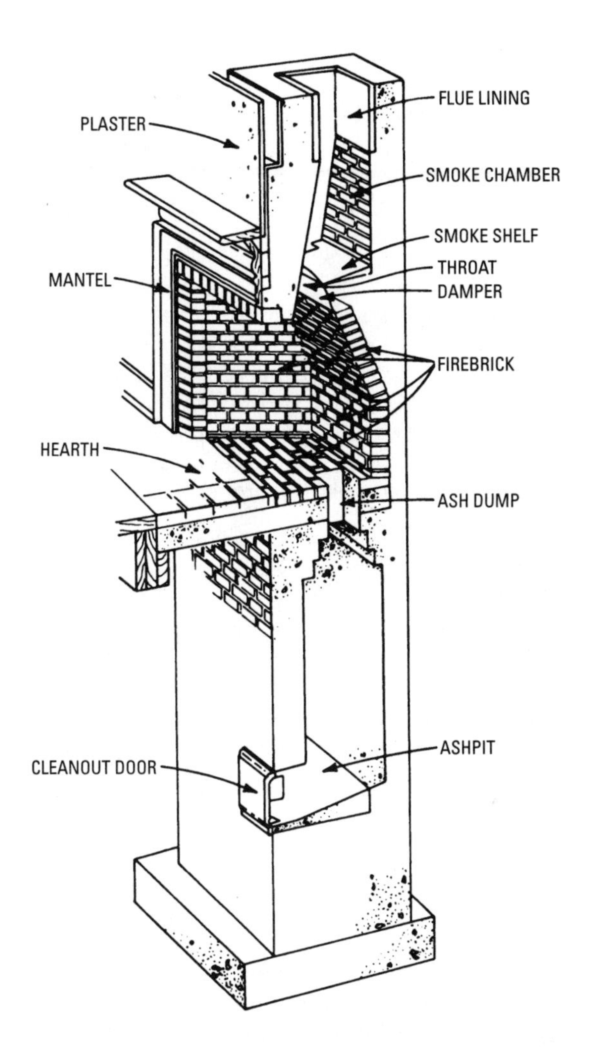

Figure 12-10 Construction details of a typical fireplace.

basement is located below the fireplace, an ash dump on the hearth near the back of the firebox is convenient. This dump consists of a metal frame about 5 inches × 8 inches with a pivoted plate through which the ashes can be dropped into a pit below.

With wooden floors, the hearth in front of the fireplace should be supported by masonry trimmer arches or other fire-resistant material, as shown in Figure 12-7 and Figure 12-10. The hearth should project at least 16 inches from the fireplace breast. It should be constructed of brick, stone, terra cotta, reinforced concrete, or other fireproof material at least 4 inches thick. The length of the hearth should not be less than the width of the fireplace opening plus 16 inches. The wood centering under the trimmer arches may be removed after the mortar has set, although it may be left in place if desired.

Wall Thickness

Fireplace walls should never be less than 8 inches thick if made of brick, or 12 inches thick if made of stone. The back and sides of the firebox may be laid up with stone or hard-burned brick, although firebrick in fire clay is preferred. When firebricks are used, they should be laid flat with their long edges exposed to lessen the danger of falling out. If they are placed on edge, however, metal ties should be built into the main brickwork to hold the firebrick in place.

Jambs

The jambs should be wide enough to give stability and a pleasing appearance to the fireplace, and are usually faced with ornamental brick, stone, or tile. When the fireplace opening is less than 3 feet wide, 12- or 16-inch jambs are usually sufficient, depending on whether a wood mantel is used or if the jambs are of exposed masonry. The edges of a wood mantel should be kept at least 8 inches away from the fireplace opening. Similar proportions of the jamb widths should be kept for wider openings and larger rooms.

Lintels

Lintels are used to support the masonry over the fireplace openings. These lintels may be $^{1}/_{2}$-inch × 3-inch bars, or $3^{1}/_{2}$-inch × $3^{1}/_{2}$-inch × $^{1}/_{4}$-inch angle iron. Heavier lintels are required for wider openings.

If a masonry arch is used over the opening, the jambs should be heavy enough to resist the thrust imposed on them by the arch. Arches used over openings less than 4 feet wide seldom sag, but wider openings will cause trouble in this respect unless special construction techniques are employed.

Throats

The sides of the fireplace should be vertical up to the throat or damper opening, as shown in Figure 12-11. The throat should be 6 or 8 inches or more above the bottom of the lintel and have an area that is not less than that of the flue. The width of the throat should be the same as the width of the fireplace opening. Starting

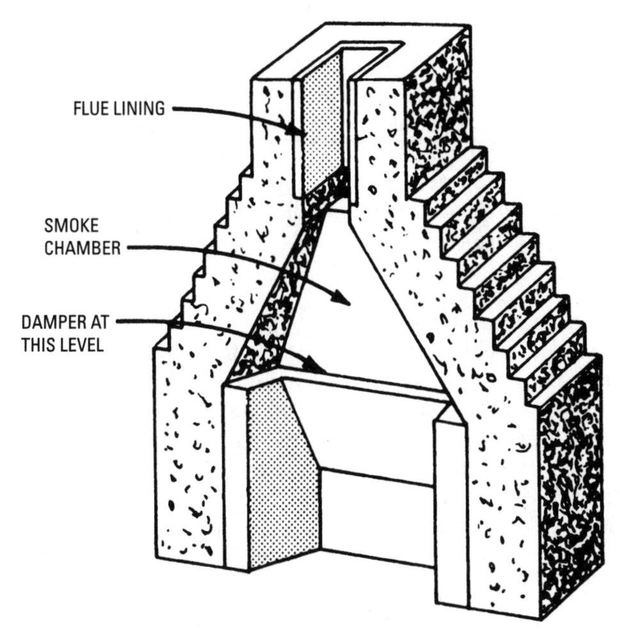

Figure 12-11 View of the fireplace and flue opening.

5 inches above the throat, the sides should be drawn in to equal the flue area.

Proper throat construction is necessary if the fireplace is to perform correctly, and the builder must ensure that the sidewalls are laid vertically until the throat has been passed and that the full length of opening is provided.

Smoke Shelf and Chamber

The smoke shelf is made by setting the brickwork back at the top of the throat to the line of the flue wall for the full length of the throat. The depth of the shelf may vary from 6 inches to 12 inches or more, depending on the depth of the fireplace.

The smoke chamber is the space extending from the top of the throat up to the bottom of the flue proper, and between the sidewalls. The walls should be drawn inward at a 30° angle to the vertical after the top of the throat is passed. These walls should be smoothly plastered with cement mortar not less than $^{1}/_{2}$-inch thick, and then coated with refractory mortar to protect the cement mortar from creosote acid damage.

Damper

A properly designed damper provides a means of regulating the draft and prevents excessive loss of heat from the room when the fireplace is not being used. A damper consists of a cast-iron frame with a lid hinged so that the width of the throat opening may be varied from completely closed to wide-open. A cap damper is mounted at the top of the chimney and is operated with a cable running down through the flue to the fireplace opening.

Flue

The area of lined flues should be at least $^1/_2$ the area of the fireplace opening, provided the chimney is at least 22 feet high, measured from the hearth. If the flue is shorter, or is unlined, its area should be at least $^1/_{10}$ the area of the fireplace opening. For example, a fireplace with an opening of 10 square feet (1440 square inches) needs a flue area of approximately 120 square inches if the chimney is 22 feet tall or 144 square inches if less than 22 feet. The nearest commercial size of rectangular lining that would be suitable is the 13-inch × 13-inch size. A 15-inch round liner would also be suitable.

Table 12-3 Sizes of Flue Linings for Various Sizes of Fireplaces*

Area of Fireplace Opening (Square Inches)	*Outside Dimensions of Standard Rectangular Flue Lining (Inches)*	*Inside Diameter of Standard Round Flue Lining (Inches)*
600	$8^1/_2 \times 8^1/_2$	10
800	$8^1/_2 \times 13$	10
1000	$8^1/_2 \times 18$	12
1200	$8^1/_2 \times 18$	12
1400	13 × 13	12
1600	13 × 13	15
1800	13 × 18	15
2000	13 × 18	15
2200	13 × 18	15
2400	18 × 18	18
2600	18 × 18	18
2800	18 × 18	18
3000	18 × 18	18

*Based on a flue area equal to one-twelfth the fireplace opening.

It is seldom possible to obtain a liner having the exact cross-sectional area required, but the inside area should never be less than that required. Always select the next larger size liner in such a case. Table 12-1 or Table 12-3 can be used to select the proper size liner or for determining the size of fireplace opening for an existing flue. The area of the fireplace opening in square inches is obtained by multiplying the width by the height, both in inches.

Summary

Proper construction of a chimney is more complicated than the average person realizes. A chimney can be too large as well as too small. Not only must it be mechanically strong, it must also be properly designed to provide the correct draft. A frequent cause of poor draft is that the peak of the roof extends higher than the top of the chimney.

A good foundation is necessary to support a properly built chimney. If a chimney is built on an outside wall, the foundation must extend well below the frost line. If a large fireplace is built in connection with the chimney, the foundation should extend at least 6 inches on all sides with reinforced concrete.

A straight flue is the most efficient design constructed. If an offset is necessary, it should not form an angle of more than 30 degrees. The most widely used material in chimney construction is brick or stone, and a chimney should never be designed without the use of a flue liner. The flue liner should always extend the entire height of the chimney.

All fireplaces should be built according to all rules and regulations to perform safely. A fireplace should always blend with the room area from the standpoint of appearance and operation.

Review Questions

1. Why should chimneys always be built in a straight line?
2. How high should a chimney be built above the roof? Why?
3. Explain the purpose of the flue lining.
4. What should be considered when designing a fireplace?
5. Explain the purpose of the hearth, throat, and damper.
6. A chimney must be designed to suit its users and blend with the ______ style of the building.
7. If the chimney is built on an outside wall, the foundation must extend to below the ____ line.

8. True or false, fireplaces are considered appropriate in living rooms, dining rooms, and bedrooms.
9. True or false, prefabricated fireplaces are available in a limited number of designs and price ranges.
10. Fireplace construction is an exacting ______.

Chapter 13

Decks

A deck is an exterior floor structure designed to extend interior living space to the outdoors (Figure 13-1). This chapter provides an overview on the functionality of decks, proper planning and layout of decks, and tips on proper construction.

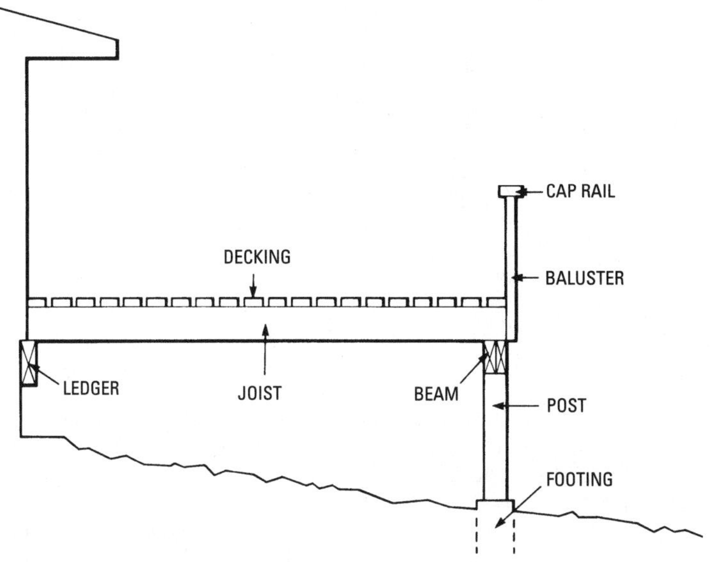

Figure 13-1 Side view of a deck. *(Courtesy of Georgia-Pacific Corp.)*

Types of Decks

One popular style of deck is the *attached, elevated* type (Figure 13-2). In this configuration, the house supports one end of the deck; posts and beams hold up the other. Attach the deck with carriage bolts through a deck beam and the siding, sheathing, and header of the house. Another method used is to bore holes in the concrete foundation of the house, insert expansion-type anchors, and attach a ledger to the house with lag screws. Then nail the deck joists to this ledger.

The *free-standing elevated* deck (Figure 13-3) is a deck that stands on its own feet, not more than two feet off the ground to the top of the decking. You can build it anywhere—adjacent to your house or in a corner of the backyard.

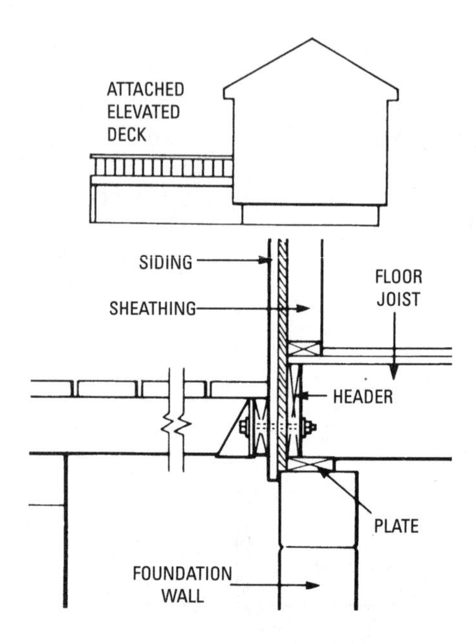

Figure 13-2 Attached elevated deck. *(Courtesy of Georgia-Pacific Corp.)*

A *bi-level* deck has the charm of two decks in one (Figure 13-4). It's the perfect way to tame a sloping lot and add living space. Bi-level deck construction amounts to building two separate decks. Then the two levels are connected with stairs.

Begin by building an elevated deck, either attached to the house or free-standing. Next set the posts for the lower-level deck. Ensure that the horizontal and vertical distances between the levels are not so great that the connecting stairs will be difficult to use. Finish off the two decks with decking of your choice. Then add the stairs and stair rails.

Planning and Layout

Before you start building decks, it's a good idea to look at all factors that might affect your plan. Check local restrictions in zoning and building codes. Is there a setback requirement specifying how far a deck must be removed from the property line? If so, you may need to call in a surveyor. Spend some time investigating the property deed and the location of underground utilities. If there's an easement for a buried water line, for example, you won't want to build a deck over that line. Contact a qualified electrician or local electrical inspector if you are unsure of the location of underground electrical lines. Do not try locating these by digging.

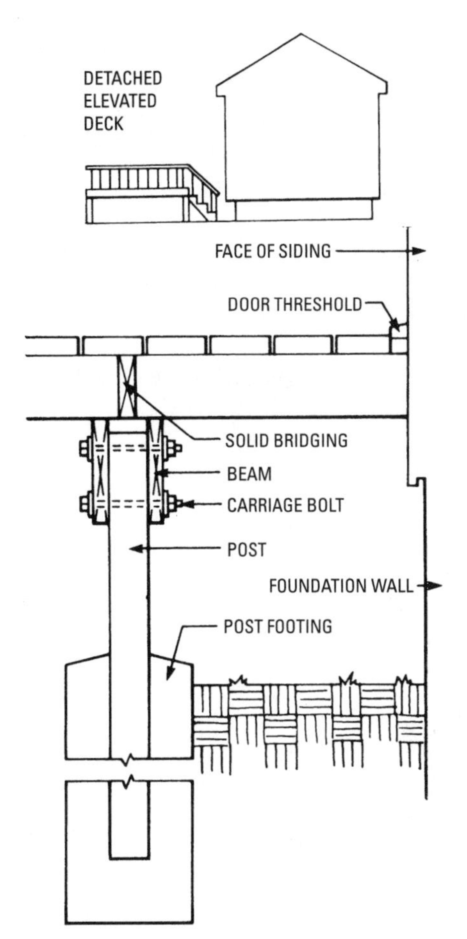

Figure 13-3 Detached deck. *(Courtesy of Georgia-Pacific Corp.)*

All good decks begin with sketches. Your sketch—even if very rough—will help you determine the best location, size, and shape of the structure.

With your sketch, a tape measure, and any necessary building permits in hand, go outside to lay out the job full-scale. Determine the width and length of the structure, measure these distances, and stake them out. Start by locating one corner adjacent to the house. Drive in a stake at that point. Measure the length of the deck and drive in a second stake adjacent to the house. From either of these stakes, measure off the width of the deck, following a line perpendicular to the wall of the house. Drive the first outside corner stake just deep enough to keep it upright. Locate the fourth corner by measuring the length and width from the appropriate stakes.

To check the corners for square, use some string and trigonometry. From a corner, measure 3 feet along the back of the deck and 4 feet along one side. If the angle between the back and the side is 90 degrees, the distance between the two points (the hypotenuse of the right triangle) will be 5 feet (Figure 13-5). This system works equally well if you measure 6 feet along the back and 8 feet along the side, obtaining a hypotenuse of 10. If the hypotenuse is too long or too short, swing the side string to the left or the right.

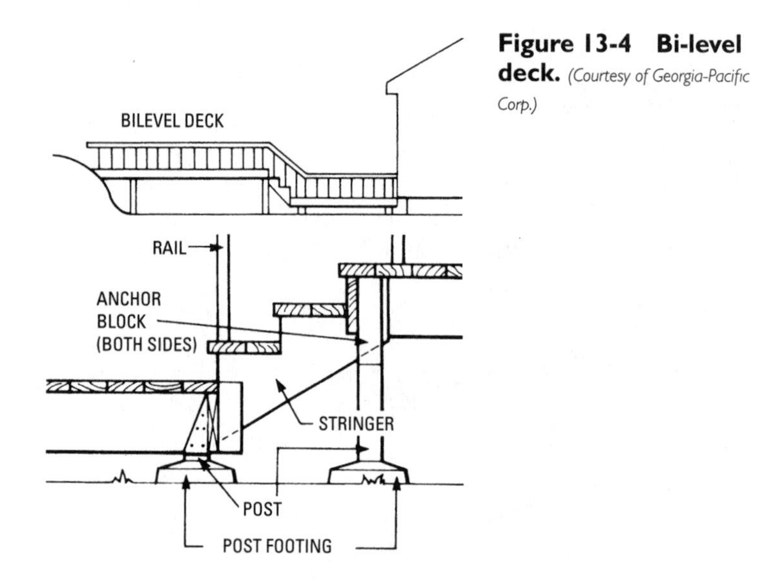

Figure 13-4 Bi-level deck. *(Courtesy of Georgia-Pacific Corp.)*

Next, measure and mark the location of the posts and drive a stake to mark the center of each one.

If you'll be securing the deck to a wall of the house, mark the location of the ledger, a strip of lumber used as a narrow ledge to support the joists on that side. Also, determine the height of the deck by measuring up the side of the house and marking where you want the top of the decking to be located.

Preparation

Proper preparation entails the correct selection of a suitable site and choosing the right materials.

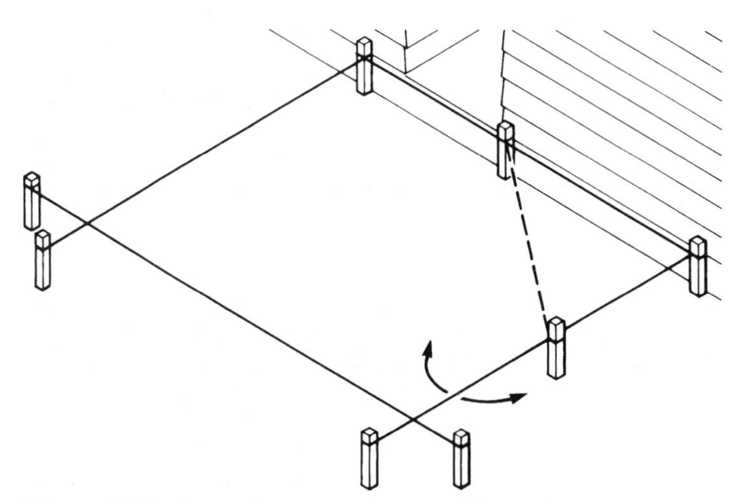

Figure 13-5 Layout. *(Courtesy of Georgia-Pacific Corp.)*

Site

Often, the area under your deck will remain shaded and, therefore, be unsuitable for grass and other ground covers. You may want to remove the sod and save it for use on bare spots elsewhere in your lawn. To prevent weeds from growing under your deck after it's finished, cover the bare soil with heavy black plastic film and top it with crushed stone, pebbles, or bark.

Next, dig the holes for the footings. These should extend below the frost line. The precise depth varies geographically, although a minimum of 2 feet is necessary.

Select Materials

A deck is only as good as the wood used to make it. For an outdoor project, the ideal lumber is that which is both economical and highly resistant to moisture and insects. Pressure-treated lumber goes through a process that forces preservatives deep into its fiber. When you see this lumber, it may still have the yellow or green color given it by those preservatives. The wood will fade to silver gray when exposed to weather. You can allow it to go gray or stain it. Some grades of redwood or cedar are also resistant to moisture and insects.

Use and maintain an effective sealer on all pressure-treated lumber projects. More information on finishing is given below. Dispose of all scraps of pressure-treated materials in ordinary trash. Do not burn.

When choosing hardware, remember that nongalvanized nails rust and bleed if exposed to rain and snow. To prevent your pressure-treated deck from becoming stained, use hot-dipped galvanized nails. Use aluminum nails with cedar and redwood.

Use and maintain an effective semi-transparent stain and water repellent on all pressure-treated lumber projects. There are many water repellents and semi-transparent stains available for exterior wood surfaces such as decks, porches, stairs, and railings that are exposed to harsh environmental conditions.

The best method of application for deck stain or a water repellent is by natural bristle brush. In either case, only one coat should be applied on smooth or rough wood. When applying the stain, it is extremely important to maintain a wet edge to avoid lap marks. *When dipping the paint brush in the can and applying it to the wood, always apply each new brush load to an unfinished area and brush toward the wet edge and then work back toward the dry edge.* Spray or roller application with a short nap lambswool roller will provide good results if followed by back brushing while the coating is still wet. This will ensure an even application.

When applying stains or water repellents to your deck, all surfaces should be clean, dry, and free of mildew, dirt, dust, oil, soot, or other contaminants. These contaminants can be removed by washing with a solution of one-quart bleach, three quarts water, and 1 cup non-ammoniated detergent followed by a thorough rinsing and drying before staining.

Deck stain or deck coatings will give maximum performance when applied to wood with a moisture content below 20 percent. However, when purchased, deck materials often have high moisture content. This is particularly true of most new pressure-treated wood. Even so, to protect the wood from the damaging ultraviolet rays of the sun, it should be coated either before or immediately after installation with a semi-transparent stain.

The wood need only be *surface dry*, not *core dry*.

When unsure about the dryness of wood, place a few drops of water on the surfaces to be stained. If the wood is *surface dry*, the water will absorb into the wood and the wood will be dry enough to apply stain or deck coatings.

An approximate coverage of 250 square feet per gallon on smooth wood and 150 square feet per gallon on rough wood can be expected. However, since less is better on some decks, as much as 400 square feet of coverage is acceptable. The temperature during application should be between 50°F and 80°F. Cooler temperatures may slow the drying process but will not harm the coating's quality or appearance. In warm or hot weather, stain or deck coatings should be applied in the shade. Stain or deck coatings will normally dry in approximately 12 hours under ideal conditions. The coating should be allowed to dry for 24 hours before replacing furniture, planter boxes, and so on.

Construction

The actual construction of the deck includes the following:

- Anchoring the posts
- Laying the beams
- Attaching the joists
- Laying the decking
- Building the stairs
- Adding deck rails
- Building stair rails

Anchoring the Posts

Because they are the cornerstones of your deck, posts deserve special attention. More than a few decks have suffered an early demise because their posts were not set properly. The preferred method of setting posts is attaching them to concrete footings. In firm soil, you can dig holes that double as forms. If you are using pressure-treated lumber, be sure it is designed for ground contact. Above ground, you'll need to build forms for concrete pedestals. Both kinds of footings protect posts from moisture, rot, and insects. Often, posts are used as a part of the railing around your deck. So, be sure not to cut off the top of the posts until you have figured which will be used in building your railings.

Of the several methods of anchoring posts to concrete, metal connectors (which raise a post up off the concrete) are best (Figure 13-6). They ensure optimal rot-resistance. Attachment methods that allow direct wood-to-wood or wood-to-concrete contact can allow moisture to seep in, possibly leading to rot.

A far less moisture-resistant method places the posts directly in wet concrete (Figure 13-7). You can buy waxed fiber forms and cut them with a handsaw. The bottom of the post must rest on 5 to 6 inches of concrete and the concrete collar must extend at least 2 inches above grade.

When setting the posts (either on top of footings or in concrete), ensure that they are vertical and aligned with one another. Use a carpenter's level to check for perpendicularity. Check each post twice,

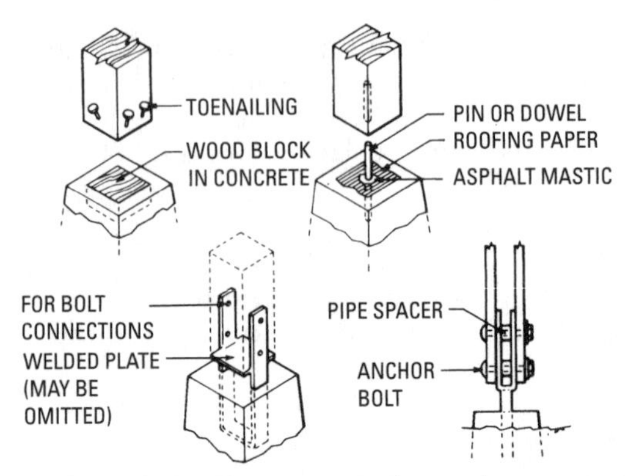

Figure 13-6 Four ways of securing posts. The two at the bottom are recommended. Note that the method shown at bottom right has a built-up column or post made of two narrow boards. Refer to the section on beams for more information about this type of post. *(Courtesy of Georgia-Pacific Corp.)*

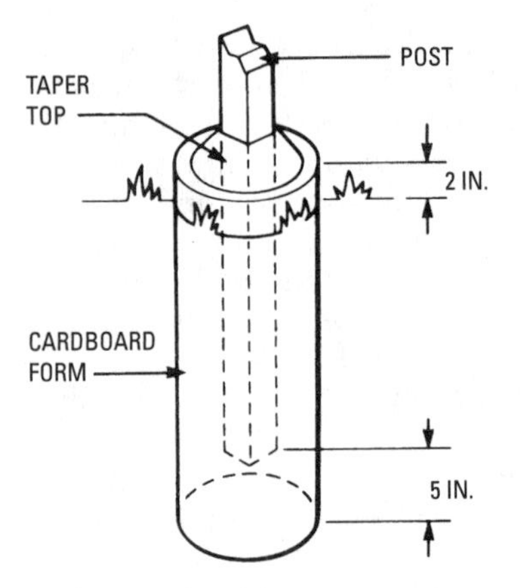

Figure 13-7 Setting wood post in wet concrete. *(Courtesy of Georgia-Pacific Corp.)*

taking readings on any two sides at right angles to each other (Figure 13-8).

Laying the Beams

At the house wall, tack one end of a leveling line where the surface of the deck will be. Pulling this line level taut, locate the same height on one of the posts. Subtract the combined thicknesses of a deck board, joist, and beam to establish the correct level on all of the posts. Saw each of them off evenly. After resting the beam on top of

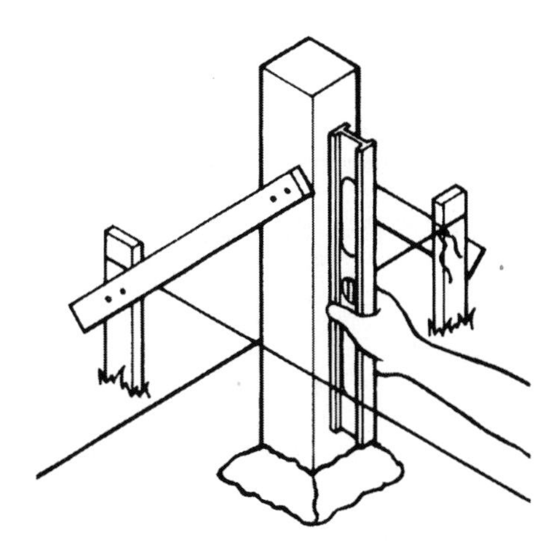

Figure 13-8 Use sturdy stakes and braces to hold footing-mounted posts plumb (vertical) during construction or to keep concrete-encased posts plumb until the concrete hardens.

(Courtesy of Georgia-Pacific Corp.)

the post, fasten them together (Figure 13-9). Joints made with lag screws or bolts are strongest. If you plan to run the posts up through the deck to serve as supports for a railing, fashion the beams of two lengths of "2 ×" materials and sandwich the posts between them. If the deck is directly on grade, cut the posts to the correct height and fasten the beams securely atop them.

Attaching the Joists

You can attach joists to beams by either resting the joists directly on top of the beams or attaching them with fasteners called *framing anchors*, or by securing them to the sides of the beams with *joist hangers*. If you use joist hangers, make sure the top of the joist is perfectly flush with that of the beam when the decking is to extend out and cover the beam. The distance between joists depends on the type and size of lumber you use. Typically, spans run from 16 to 32 inches. Check with your local building supply retailer to determine the appropriate span between joists for your project.

Nail solid bridging (intermediate joists) between the main joists when spanning long distances (Figure 13-10).

Laying the Decking

Decking is most often made of 2 × 2 lumber, 2 × 4 lumber, and 2 × 6 lumber. When you start nailing the decking to the joists, you begin to see how enjoyable your outdoor living area will be!

One of the easiest methods of laying decking is running it all in one direction. With a little more work and a bit more material, you can create intricate patterns (Figure 13-11 and Figure 13-12). If you

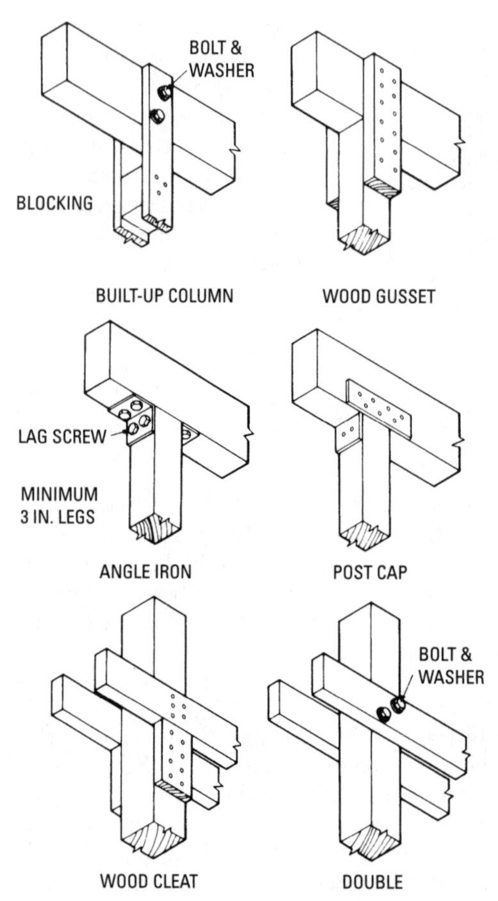

Figure 13-9 Methods for attaching posts to beams. *(Courtesy of Georgia-Pacific Corp.)*

choose one of these or design your own pattern, be sure to build in adequate joist supports for all the decking boards.

Be sure to lay the first board accurately, because it will serve as a guide for the rest of the deck surface. Don't worry about cutting the decking to the exact length yet—you'll be able to do that after all the decking boards are in place.

If the lumber you're using is slightly warped the long way, nail one end of the board to a joist, then pull the unattached end into position and nail it, working from joist to joist. Then drive two nails

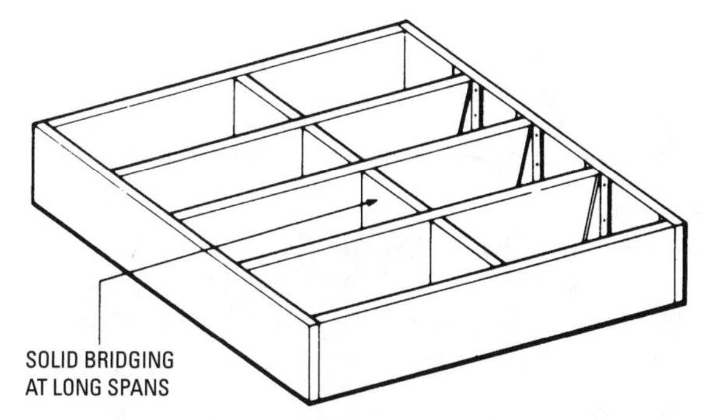

Figure 13-10 Bridging will keep the joists from twisting and help even out the load on the deck. *(Courtesy of Georgia-Pacific Corp.)*

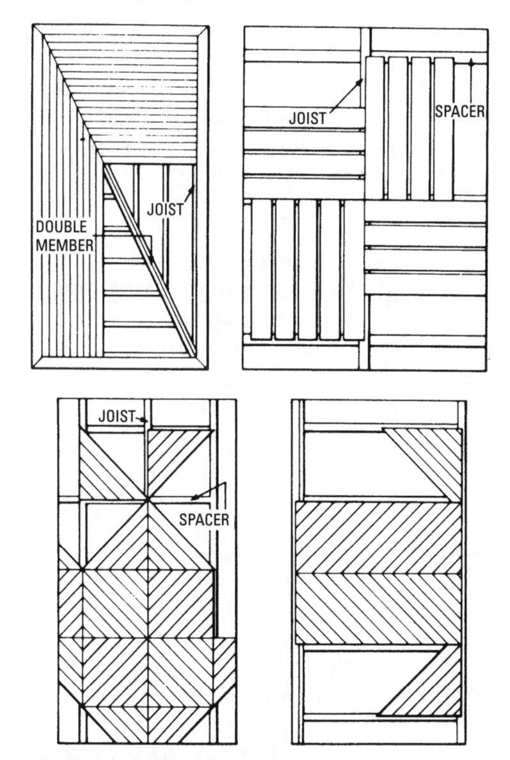

Figure 13-11 Decking with a 90° pattern, a basket weave, a harlequin, and a herringbone. *(Courtesy of Georgia-Pacific Corp.)*

Figure 13-12 Always check the end grain of each board before nailing the board in place. Notice that the tree rings curve toward one side of the board, the bark side. Nail all decking bark side up to minimize cupping. *(Courtesy of Georgia-Pacific Corp.)*

into each decking board at each joist. To help prevent the decking from splitting as you nail it in place (especially at the ends of the boards), either blunt the point of each nail slightly with a hammer before driving it or drill pilot holes for the nails. Stagger the end-to-end joints as you go across the deck.

From time to time, stop to measure the distance remaining to be covered with decking. This will help you keep the decking boards parallel (Figure 13-13). If some boards are uneven, make any corrections gradually. Maintain proper spacing between decking boards with a $^1/_4$-inch spacer strip (Figure 13-14).

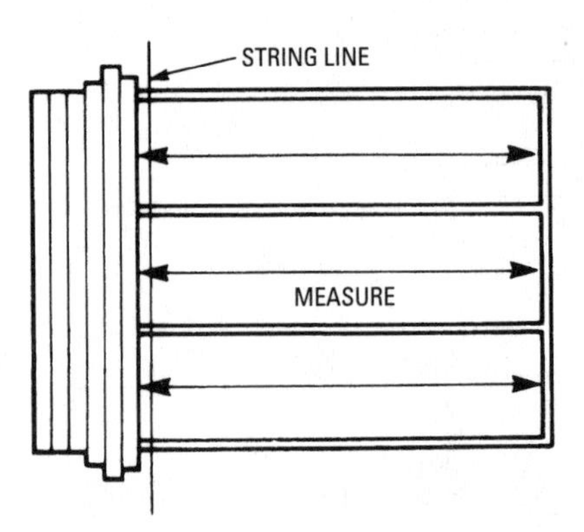

Figure 13-13 Take measurements along both sides and the middle to ensure even spacing. *(Courtesy of Georgia-Pacific Corp.)*

When you have about 6 feet of decking left to position, begin adjusting the spacing between boards to avoid an unsightly gap at the end of the deck.

After all the boards are nailed, snap a chalk line along the edges and nail on a board as a guide for cutting and trim the edges with a saw (Figure 13-15).

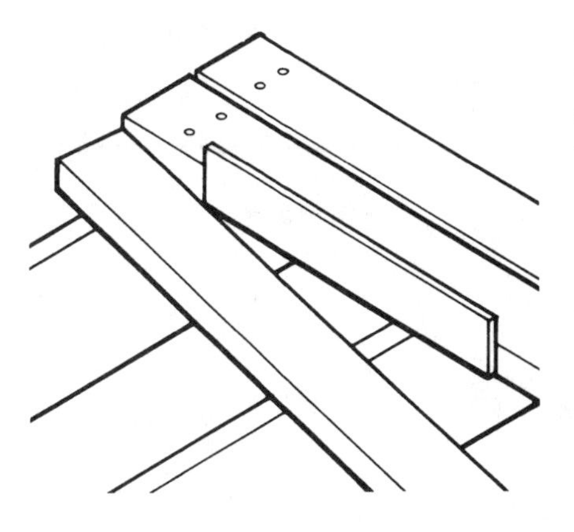

Figure 13-14 Use a thin strip of wood as a spacer. *(Courtesy of Georgia-Pacific Corp.)*

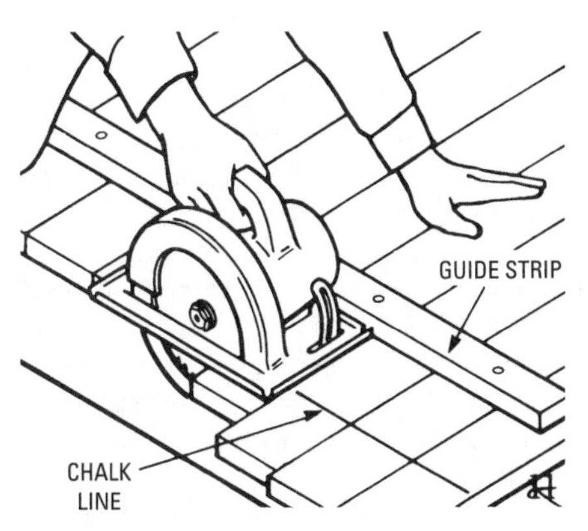

Figure 13-15 Using a guide strip to provide a very even saw cut.
(Courtesy of Georgia-Pacific Corp.)

Building the Stairs

Laying out a stairway elevates you from the rank of beginner. Success depends on how accurately you compute angles, and maintain equal riser and tread spacing from top to bottom. Time and care will help you meet the challenge. You may want to start with the open-riser/open-stringer stairway, which it is the easiest to build. Don't forget that you can also purchase prefabricated stairs and railings. Check with your local retailer for availability.

The *stringers* (or side supports of the treads) are the most important part of your stairs. They must be accurately cut for the sake of safety and appearance, and strong enough to carry heavy loads.

Think of stairs as a means of dividing a difference of elevation into a series of equal steps. Measure the height from the deck top to the ground (the *total rise*) and the distance from the deck to where the stairs will end (the *total run*), as shown in Figure 13-16.

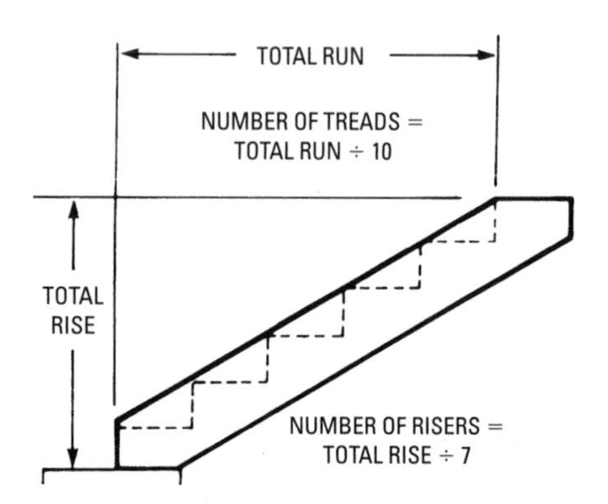

Figure 13-16 Figuring the total rise and run of the stairs. *(Courtesy of Georgia-Pacific Corp.)*

Keep in mind this general guideline: Most steps have a rise of about 7 inches and a run of 10 to 11 inches. Divide those figures into the rise and run to see if you can cover the distance with an even number of steps. If you can't do so at first, adjust the rise and run measurements by equal amounts until you do come out even.

Don't forget to include the thickness of the tread material in your rise calculations. A common choice for treads is "2 ×" lumbers. Depending on your tread dimension you can use a 2 × 10, a 2 × 12, or two 2 × 6 boards spaced 1/4-inch apart. For an even stronger step, consider building treads of 2 × 4 boards set on edge. Separate the boards with 3/8-inch plywood spacers.

Mark off the rise height on one leg of a framing square (an L-shaped ruler) and the depth of the tread on the other (Figure 13-17). Lay the square against a 2 × 10 or 2 × 12 and mark the lumber. Subtract the tread thickness from the bottom; add it to the top (Figure 13-18).

Cut along the lines, and then use this stringer as a pattern for the others (Figure 13-19). When cutting, be sure to compensate for the thickness of the saw blade.

Fasten the stringers to the sides of a joist or header using a metal hanger, or bolt the steps to the joists (Figure 13-20). Next, nail on the treads (bark side up), using large hot-dipped galvanized nails. Then trim the treads flush with the stringer or let them overlap (Figure 13-21).

Set the bottom of the steps on a concrete footing and secure it with angle irons, or lag screws and fasteners called *expansion anchors* (Figure 13-22).

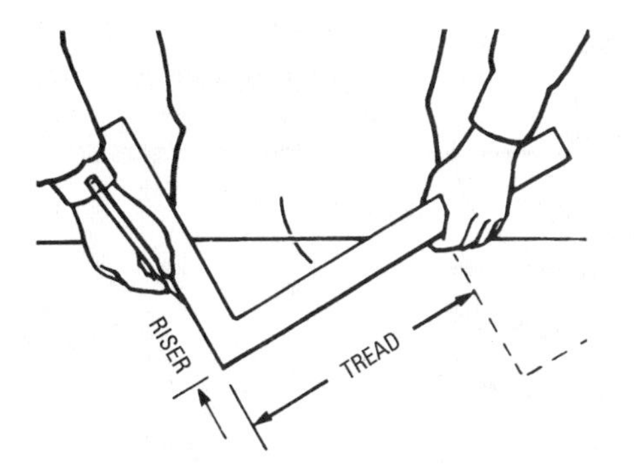

Figure 13-17 Laying out the stringer. *(Courtesy of Georgia-Pacific Corp.)*

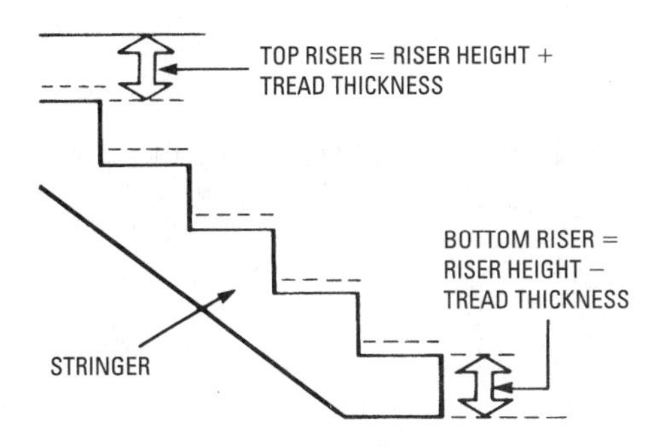

Figure 13-18 Figuring the top and bottom riser. *(Courtesy of Georgia-Pacific Corp.)*

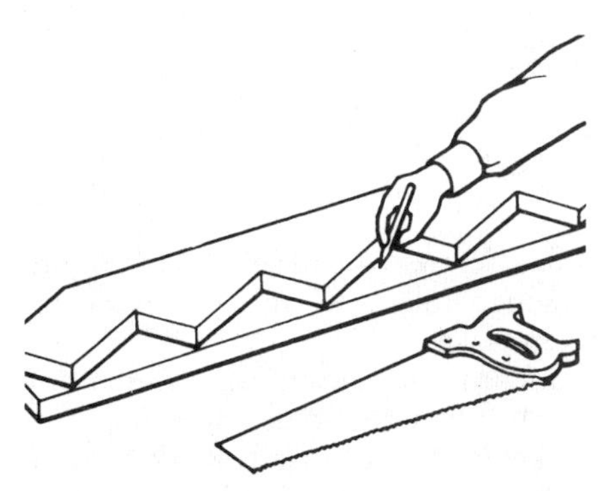

Figure 13-19 Use the first stringer as a pattern for the others. *(Courtesy of Georgia-Pacific Corp.)*

Adding Deck Rails

Here's some room for creative design. Just remember to make your railings strong. The chief purpose of a railing is to provide safety for persons on the deck. For that reason, space the vertical and horizontal members closely enough to keep children from slipping through the gaps (Figure 13-23). Also, secure the railing posts to

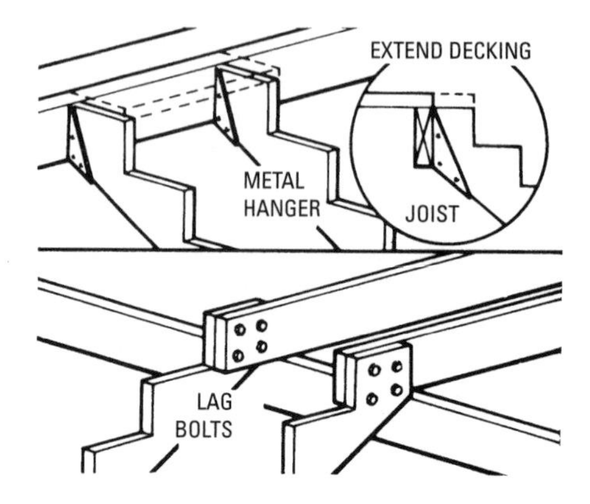

Figure 13-20 Attaching stringers at the top. *(Courtesy of Georgia-Pacific Corp.)*

Figure 13-21 Nail on the treads. *(Courtesy of Georgia-Pacific Corp.)*

the deck understructure with lag screws or carriage bolts (Figure 13-24). Nails do not provide enough holding power. Do not rely on the decking to hold the railing in place.

Figure 13-25 shows five standard types of railings. The first is made of 4 × 4 posts and 2 × 4 rails. Other types of railings are variations of the traditional picket fence. Their framework consists of 2 × 4 or 4 × 4 posts spanned by horizontal 2 × 4 rails. The trick to keeping pickets aligned is to hold a spacer in place while nailing up each one. Make sure the pickets are vertical by occasionally checking with a carpenter's level. To prevent these styles from sagging, keep spans less than 6 feet.

The next two types of railings shown are made of evenly spaced vertical boards (2 × 2 or 1 × 8). They form railings that provide

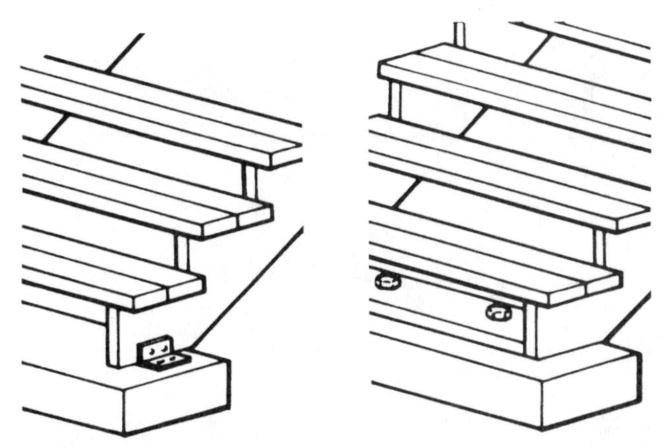

Figure 13-22 Attaching stringers at the bottom. *(Courtesy of Georgia-Pacific Corp.)*

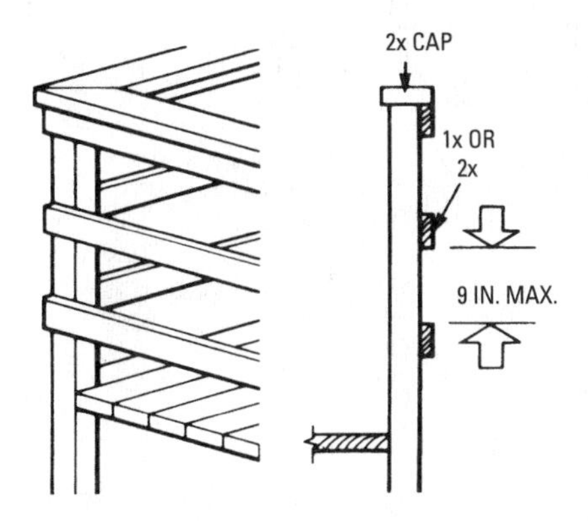

Figure 13-23 Horizontal rails. *(Courtesy of Georgia-Pacific Corp.)*

some privacy, yet permit free air movement. For still more privacy, or to screen out unsightly scenery, use 2-foot square panels spaced 4 inches apart. You may want to consider lumber siding or grooved plywood panels. They enclose a railing completely, transforming it into a low wall open only at the bottom. For wind control, the side-by-side panels offer best protection.

A handsome and useful addition to many decks is the combination railing and seat (Figure 13-26). For standard seating, build benches 15 to 18 inches high, and be sure the seats are at least 15 inches wide. Backrests should extend at least 12 inches above the seats.

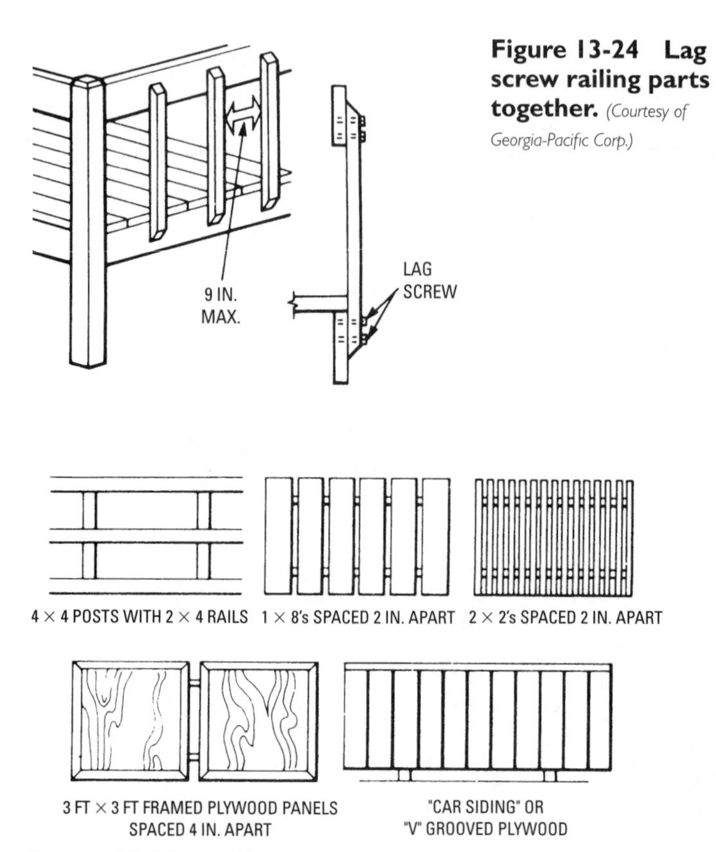

Figure 13-24 Lag screw railing parts together. *(Courtesy of Georgia-Pacific Corp.)*

Figure 13-25 Railing types. *(Courtesy of Georgia-Pacific Corp.)*

Building Stair Rails

Stair rails perform the all-important function of safeguarding against accidents. Even a short stair should have a rail 30 to 40 inches tall to provide adequate protection. Local building codes may dictate minimum requirements for your stair rails.

The key elements of a stair rail are the posts. You may use balusters as shown in Figure 13-27, a good safety feature, especially if children use the stairs often, or simply one post at either end of the rail. If you plan to use a 2 × 4 top rail, keep the span between posts under 6 feet. With 2 × 6 or larger rails, you can allow spans of up to 8 feet. Attach the posts to the stair stringers or deck framing with bolts or lag screws.

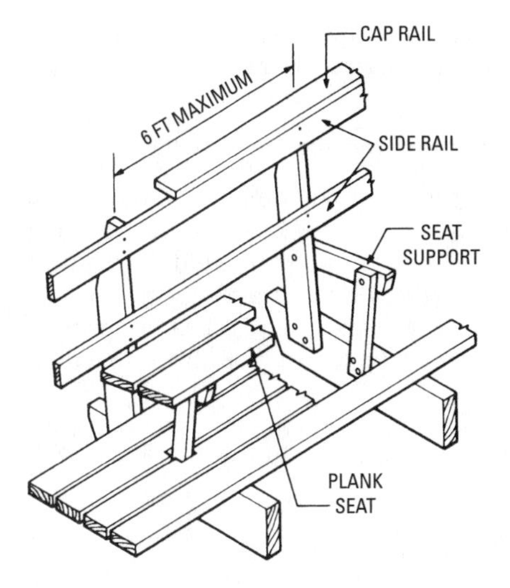

Figure 13-26 Railing and seat combination.
(Courtesy of Georgia-Pacific Corp.)

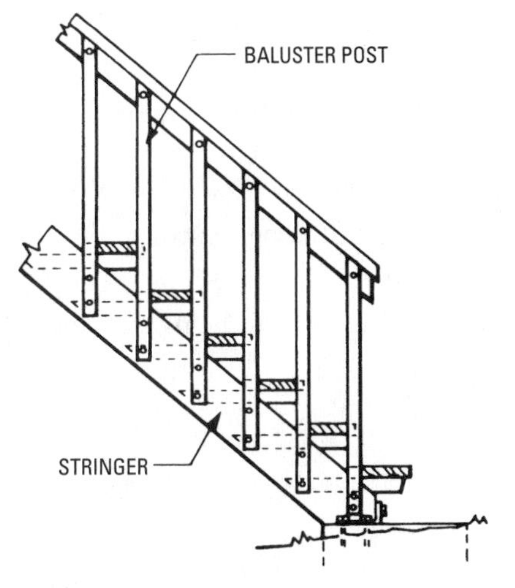

Figure 13-27 Stair railing. *(Courtesy of Georgia-Pacific Corp.)*

Summary

Decks are exterior floor structures that extend living space to the outdoors. There are three main types of construction: attached, elevated, and bi-level.

When planning a deck, research zoning and site conditions, and then set the design down on paper. Lay out the deck full-scale on the site with stakes and string.

Prepare the site and set footings into the ground. Use lumber that is resistant to moisture, decay, and insects. Protect the wood with water-repellent preservative or stain.

In deck construction, anchor the posts to concrete footings. Beams are horizontal supports for the joists and should be lag-screwed or bolted to the posts. Joists are laid directly on the beams or are fastened with framing anchors or joist hangers. Lay decking boards with the bark side of the board up. Stair building requires considerable forethought and skill. The tread and risers must be spaced carefully for safe use. Railings make high decks safe for people using the deck. Railing parts must be screwed or bolted together.

Review Questions

1. How are attached decks fastened to the house?
2. What part of a deck holds the beams up off the footings?
3. Describe the method for laying out a deck on the site.
4. What kinds of wood are best for deck construction?
5. How can you test wood to see if it is dry enough to stain or treat?
6. What type of fasteners have the strength needed in deck construction?
7. Why would you use solid bridging between joists?
8. What methods are used to keep the decking boards parallel?
9. Why are railings needed on decks and stairs?
10. True or false, one of the easiest methods of laying decking is running it all in one direction.

Appendix A

References

This appendix provides a guide to important sources of information.

Web Site Resources

Web Site	*Description*
www.acronymfinder.com	Association acronyms
www.advancedbuildings.org	Technologies for commercial buildings
www.aecinfo.com	Architecture, engineering, and construction
www.afsonl.com	Building product information
www.americanarchitecture.com	Homepage listings for the building industry
www.arcat.com/divs	Organizations by C51 divisions
www.building.org	Building industry exchange
www.buildingonline.com	Architects, contractors, and manufacturers
www.codesourcepc.com	Code-related Web sites
www.construction.com	Construction industry news
www.constructioneducation.com	Construction education
www.constructionmaterials.com	Manufacturers by C51 divisions
www.eren.doe.gov	Insulation information sources
www.floorbiz.com	Flooring product manufacturers
www.geoforum.com	Geotechnical and deep foundation information
www.homelighting.com	Lighting industry links
www.hvacmall.com	HVAC industry links
www.ipl.org	Internet Public Library general reference site
www.nssn.org	Global codes and standards
www.plumbinglinks.com	Plumbing industry links
www.mnconcretecouncil.org	Concrete industry links and information
www.solidwaste.com	Construction waste
www.thebluebook.com	Building products directory
www.virtualengineer.com	Engineering links

Professional and Trade Associations

Organization	*Address*	*Web Site*
American Arbitration Association	355 Madison Avenue, 10th Floor, New York, NY 10017-4605	www.adr.org
American Institute of Architects	17555 New York Avenue NW, Washington, DC 20006	www.e-archtect.com www.aiaonline.com
American Institute of Building Design	991 Post Road East, Westport, CN 06680	www.aibd.org
American Insurance Association	1130 Connecticut Avenue NW, Suite 1000, Washington, DC 20036	www.aiadc.org
American Society of Civil Engineers	1801 Alexander Bell Drive, Reston, VA 20191-4400	www.asce.org
American Society of Interior Designers	608 Massachusetts Avenue NE, Washington, DC 20002	www.interiors.org
American Society of Landscape Architects	656 Eye Street NW, Washington, DC 20001-5756	www.asla.org
Associated General Contractors of America	555 John Carlyle Street, Suite 200, Alexandria, VA 22514	www.agc.org
Building Research Establishment	Bucknalls Lane, Garston, Watford, WD27JR, England	www.bre.co.uk
Building Seismic Safety Council	National Institute of Building Sciences, 1090 Vermont Avenue NW, Suite 700, Washington, DC 20005-4905	www.bssconline.org
Canada Mortgage and Housing Corporation	700 Montreal Road, Ottawa, ON, K1AOP7, Canada	www.cmhc-schl.gc.ca

Organization	Address	Web Site
Canadian Construction Association	75 Albert Street, Suite 400, Ottawa, ON, K1P5E7, Canada	www.cca-acc.com
Civil Engineering Research Foundation	1015 15th Street NW, Suite 600, Washington, DC 20005-2605	www.cerf.org
Construction Management Association of America	7918 Jones Branch Drive, Suite 540, McLean, VA 22102	www.cmaanet.org
Construction Specifications Canada	100 Lombard Street, Suite 200, Toronto, ON, M5C 1M5, Canada	www.csc-dcc.ca
Construction Specifications Institute	99 Canal Center Plaza, Suite 500, Alexandria, VA 22514-1791	www.csinet.org
Energy Efficient Building Association	10740 Lyndale Avenue South, Suite 10W, St. Paul, MN 55420-5615	www.eeba.org
Environmental Protection Agency	1200 Pennsylvania Avenue NW, Washington, DC 20460	www.epa.gov
Green Building Information Council	150 Lewis Street, Ottawa, ON, K2POS7, Canada	www.greenbuilding.ca

Glossary

Air-dried lumber—Lumber that has been piled in yards or sheds for any length of time. For the United States as a whole, the minimum moisture content of thoroughly air-dried lumber is 12 to 15 percent, with the average somewhat higher.

Airway—A space between roof insulation and roof boards for movement of air.

Alligatoring—Coarse checking pattern characterized by a slipping of the new paint coating over the old coating to the extent that the old coating can be seen through the fissures.

Anchor bolts—Bolts to secure a wooden sill to a concrete or masonry floor, foundation, or wall.

Apron—The flat member of the inside trim of a window placed against the wall immediately beneath the stool.

Areaway—An open subsurface space adjacent to a building used to admit light or air, or as a means of access to a basement or cellar.

Asphalt—Most asphalt is a residue from evaporated petroleum. It is insoluble in water, but is soluble in gasoline and melts when heated. Used in making waterproof roof coverings of many types, exterior wall coverings, flooring tile, and the like.

Astragal—A molding, attached to one of a pair of swinging doors, against which the other door strikes.

Attic ventilators—In home building, usually openings in gables or ventilators in the roof. Also, mechanical devices to force ventilation by the use of power-driven fans. Also see *Louver*.

Backband—Molding used on the side of a door or window casing for ornamentation or to increase the width of the trim.

Backfill—The replacement of excavated earth into a pit or trench or against a foundation wall.

Balusters—Small spindles or members forming the main part of a railing for a stairway or balcony, fastened between a bottom and top rail.

Base or baseboard—A board placed against the wall around a room next to the floor to provide a proper finish between the floor and plaster.

Base molding—Molding used to trim the upper edge of interior baseboard.

Base shoe—Molding used next to the floor on interior baseboard. Sometimes called a *carpet strip*.

Batten—Narrow strips of wood or metal used to cover joints.

Batter board—One of a pair of horizontal boards nailed to posts set at the corners of an excavation, used to indicate the desired level. Also, a fastening for stretched strings to indicate the outlines of foundation walls.

Bay window—Any window space projecting outward from the walls of a building, either square or polygonal in plan.

Beam—A structural member transversely supporting a load.

Bearing partition—A partition that supports any vertical load in addition to its own weight.

Bearing wall—A wall that supports any vertical load in addition to its own weight.

Bed molding—A molding set in an angle, as between an overhanging cornice or eaves, of a building and the side walls.

Blinds (shutters)—Light wood sections in the form of doors to close over windows to shut out light, give protection, or add temporary insulation. Commonly used for ornamental purposes, in which case they are fastened rigidly to the building.

Blind-nailing—Nailing in such a way that the nailheads are not visible on the face of the work.

Blind stop—A rectangular molding, usually 3/4 by 1 3/8 inches or more, used in the assembly of a window frame.

Blue stain—A bluish or grayish discoloration of the sapwood caused by the growth of certain mold-like fungi on the surface and in the interior of the piece, made possible by the same conditions that favor the growth of other fungi.

Bolster—A short horizontal timber resting on the top of a column for the support of beams or girders.

Boston ridge—A method of applying asphalt or wood shingles as a finish at the ridge or hips of a roof.

Brace—An inclined piece of framing lumber used to complete a triangle, and thereby to stiffen a structure.

Brick veneer—A facing of brick laid against frame or tile wall construction.

Bridging—Small wood or metal members that are installed in a diagonal position between the floor joists to act both as tension and compression members for the purpose of bracing the joists and spreading the action of loads.

Buck—Often used in reference to rough frame opening members. Door bucks are used in reference to a metal door frame.

Built-up roof—A roofing composed of three to five layers of rag felt or jute saturated with coal tar, pitch, or asphalt. The top is finished with crushed slag or gravel. Generally used on flat or low-pitched roofs.

Butt joint—The junction where the ends of two timbers or other members meet in a square-cut joint.

Cabinet—A shop- or job-built unit for kitchens. Cabinets often include combinations of drawers, doors, and the like.

Cant strip—A wedge- or triangular-shaped piece of lumber used at the gable ends under the shingles, or at the junction of the house and a flat deck under the roofing.

Cap—The upper member of a column, pilaster, door cornice, molding, and the like.

Casement frames and sash—Frames of wood or metal enclosing part or all of the sash that may be opened by means of hinges affixed to the vertical edges.

Casing—Wide molding of various widths and thicknesses used to trim door and window openings.

Checking—Cracks that appear with age in many exterior paint coatings, at first superficial, but which in time may penetrate entirely through the coating.

Checkrails—Meeting rails sufficiently thicker than a window to fill the opening between the top and bottom sash made by the parting stop in the frame. They are usually beveled.

Collar beam—A beam connecting pairs of opposite rafters above the attic floor.

Column—In architecture, a perpendicular supporting member, circular or rectangular in section, usually consisting of a base, shaft, and capital. In engineering, a structural compression member, usually vertical, supporting loads acting on or near and in the direction of its longitudinal axis.

Concrete, plain—Concrete without reinforcement, or reinforced only for shrinkage or temperature changes.

Condensation—Beads or drops of water (and frequently frost in extremely cold weather) that accumulate on the inside of the exterior covering of a building when warm, moisture-laden air from the interior reaches a point where the temperature no longer permits the air to sustain the moisture it holds. Use of louvers or attic ventilators will reduce moisture condensation in attics.

Conduit, electrical—A pipe (usually metal) in which wire is installed.

Construction, drywall—A type of construction in which the interior wall finish is applied in a dry condition, generally in the form of sheet materials, as contrasted to plaster.

Construction, frame—A type of construction in which the structural parts are of wood or dependent upon a wood frame for support. In codes, if brick or other incombustible material is applied to the exterior walls, the classification of this type of construction is usually unchanged.

Coped joint—See *Scribing*.

Corbel out—To build out one or more courses of brick or stone from the face of a wall to form a support for timbers.

Corner bead—A strip of galvanized iron, sometimes combined with a strip of metal lath, which is placed on corners before plastering to reinforce them. Also, a strip of wood finish three-quarters round or angular placed over a plastered corner for protection.

Corner boards—Used as trim for the external corners of a house or other frame structure against which the ends of the siding are finished.

Corner braces—Diagonal braces let into studs to reinforce corners of frame structures.

Cornerite—Metal-mesh lath cut into strips and bent to a right angle. Used in interior corners of walls and ceilings on lath to prevent cracks in plastering.

Cornice—A decorative element made up of molded members usually placed at or near the top of an exterior or interior wall.

Cornice return—That portion of the cornice that returns on the gable end of a house.

Counterflashing—A flashing usually used on chimneys at the roofline to cover shingle flashing and to prevent moisture entry.

Cove molding—A three-sided molding with concave face used wherever small angles are to be covered.

Crawl space—A shallow space below the living quarters of a house. It is generally not excavated or paved, and is often enclosed for appearance by a skirting or facing material.

Cricket—A small drainage-diverting roof structure of single or double slope placed at the junction of larger surfaces that meet at an angle.

Crown molding—A molding used on a cornice or wherever a large angle is to be covered.

d—See *Penny*.

Dado—A rectangular groove in a board or plank. In interior decoration, a special type of wall treatment.

Decay—Disintegration of wood or other substance through the action of fungi.

Deck paint—An enamel with a high degree of resistance to mechanical wear, for use on such surfaces as porch floors.

Density—The mass of substance in a unit volume. When expressed in the metric system, it is numerically equal to the specific gravity of the same substance.

Dimension—See *Lumber, Dimension*.

Direct nailing—To nail perpendicular to the initial surface or to the junction of the pieces joined. Also termed *face nailing*.

Doorjamb, interior—The surrounding case into which and out of which a door closes and opens. It consists of two upright pieces (called *jambs*) and a head, fitted together and rabbeted.

Dormer—An internal recess, the framing of which projects from a sloping roof.

Downspout—A pipe (usually of metal) for carrying rain water from roof gutters.

Dressed and matched (tongue-and-groove)—Boards or planks machined in such a manner that there is a groove on one edge and a corresponding tongue on the other.

Drier, paint—Usually oil-soluble soaps of such metals as lead, manganese, or cobalt, which, in small proportions, hasten the oxidation and hardening (drying) of the drying oils in paints.

Drip—(*a*) A member of a cornice or other horizontal exterior-finish course that has a projection beyond the other parts for throwing off water. (*b*) A groove in the under side of a sill to cause water to drop off on the outer edge, instead of drawing back and running down the face of the building.

Drip cap—A molding placed on the exterior top side of a door or window to cause water to drip beyond the outside of the frame.

Ducts—In a house, usually round or rectangular metal pipes for distributing warm air from the heating plant to rooms, or air from a conditioning device. Ducts are also made of asbestos and composition materials.

Eaves—The margin or lower part of a roof projecting over the wall.

Expansion joint—A bituminous fiber strip used to separate blocks or units of concrete to prevent cracking caused by expansion as a result of temperature changes.

Fascia (or facia)—A flat board, band, or face, used sometimes by itself, but usually in combination with moldings, located at the outer face of the cornice.

Filler (wood)—A heavily pigmented preparation used for filling and leveling off the pores in open-pored woods.

Fire-resistive—In the absence of a specific ruling by the authority having jurisdiction, applies to materials for construction not combustible in the temperatures of ordinary fires and that will withstand such fires without serious impairment of their usefulness for at least 1 hour.

Fire-retardant chemical—A chemical or preparation of chemicals used to reduce flammability or to retard spread of flame.

Fire stop—A solid, tight closure of a concealed space, placed to prevent the spread of fire and smoke through such a space.

Flagstone (flagging or flags)—Flat stones, from 1 to 4 inches thick, used for rustic walks, steps, floors, and the like. Usually sold by the ton.

Flashing—Sheet metal or other material used in roof and wall construction to protect a building from seepage of water.

Flat paint—An interior paint that contains a high proportion of pigment and dries to a flat or lusterless finish.

Flue—The space or passage in a chimney through which smoke, gas, or fumes ascend. Each passage is called a flue, which, together with any others and the surrounding masonry, makes up the chimney.

Flue lining—Fireclay or terra-cotta pipe, round or square, usually made in all of the ordinary flue sizes and in 2-foot lengths, used for the inner lining of chimneys with the brick or masonry work around the outside. Flue lining should run from the concrete footing to the top of the chimney cap. Figure a foot of flue lining for each foot of chimney.

Footing—The spreading course or courses at the base or bottom of a foundation wall, pier, or column.

Foundation—The supporting portion of a structure below the first-floor construction, or below grade, including the footings.

Framing, balloon—A system of framing a building in which all vertical structural elements of the bearing walls and partitions consist of single pieces extending from the top of the soleplate to the roofplate, and to which all floor joists are fastened.

Framing, platform—A system of framing a building in which floor joists of each story rest on the top plates of the story below or on the foundation sill for the first story, and the bearing walls and partitions rest on the subfloor of each story.

Frieze—Any sculptured or ornamental band in a building. Also, the horizontal member of a cornice set vertically against the wall.

Frostline—The depth of frost penetration in soil. This depth varies in different parts of the country. Footings should be placed below this depth to prevent movement.

Fungi, wood—Microscopic plants that live in damp wood and cause mold, stain, and decay.

Fungicide—A chemical that is poisonous to fungi.

Furring—Strips of wood or metal applied to a wall or other surface to even it, to form an air space, or to give the wall an appearance of greater thickness.

Gable—That portion of a wall contained between the slope of a single-sloped roof and a line projected horizontally through the lowest elevation of the roof construction.

Gable end—An end wall having a gable.

Gloss (paint or enamel)—A paint that contains a relatively low proportion of pigment and dries to a sheen or luster.

Gloss enamel—A finishing material made of varnish and sufficient pigments to provide opacity and color, but little or no pigment of low opacity. Such an enamel forms a hard coating that has a maximum smoothness of surface and a high degree of gloss.

Girder—A large or principal beam used to support concentrated loads at isolated points along its length.

Grain—The direction, size, arrangement, appearance, or quality of the fibers in wood.

Grain, edge (vertical)—Edge-grain lumber has been sawed parallel to the pith of the log and approximately at right angles to the growth rings (that is, the rings form an angle of 45° or more with the surface of the piece).

Grain, flat—Flat-grain lumber has been sawed parallel to the pith of the log and approximately at right angles to the growth rings (that is, the rings form an angle of less than 45° with the surface of the piece).

Grain, quartersawn—Another term for edge grain.

Grounds—Strips of wood, of the same thickness as the lath and plaster, that are attached to walls before the plastering is done. Used around windows, doors, and other openings as a plaster stop, and in other places for the purpose of attaching baseboards or other trim.

Grout—Mortar made of such consistency by the addition of water that it will just flow into the joints and cavities of the masonry work and fill them solid.

Gutter—A shallow channel or conduit of metal, wood, or plastic set below and along the eaves of a house to catch and carry off rain water from the roof.

Gypsum plaster—Gypsum formulated to be used with the addition of sand and water for base-coat plaster.

Header—(*a*) A beam placed perpendicular to joists and to which joists are nailed in framing for chimney, stairway, or other opening. (*b*) A wood lintel.

Hearth—The floor of a fireplace, usually made of brick, tile, or stone.

Heartwood—The wood extending from the pith to the sapwood, the cells of which no longer participate in the life processes of the tree.

Hip—The external angle formed by the meeting of two sloping sides of a roof.

Hip roof—A roof that rises by inclined planes from all four sides of a building.

Humidifier—A device designed to discharge water vapor into a confined space for the purpose of increasing or maintaining the relative humidity in an enclosure.

I-beam—A steel beam with a cross-section resembling the letter "I."

Insulation, building—Any material high in resistance to heat transmission that, when placed in the walls, ceilings, or floor of a structure, will reduce the rate of heat flow.

Jack rafter—A rafter that spans the distance from the wallplate to a hip, or from a valley to a ridge.

Jamb—The side post or lining of a doorway, window, or other opening.

Joint—The space between the adjacent surfaces of two members or components joined and held together by nails, glue, cement, mortar, or other means.

Joint cement—A powder that is usually mixed with water and used for joint treatment in gypsum-wallboard finish. Often called *spackle*.

Joist—One of a series of parallel beams used to support floor and ceiling loads, and supported in turn by larger beams, girders, or bearing walls.

Knot—That portion of a branch or limb that has become incorporated in the body of a tree.

Landing—A platform between flights of stairs or at the termination of a flight of stairs.

Lath—A building material of wood, metal, gypsum, or insulating board that is fastened to the frame of a building to act as a plaster base.

Lattice—An assemblage of wood or metal strips, rods, or bars made by crossing them to form a network.

Leader—See *Downspout*.

Ledger strip—A strip of lumber nailed along the bottom of the side of a girder on which joists rest.

Light—Space in a window sash for a single pane of glass. Also, a pane of glass.

Lintel—A horizontal structural member that supports the load over an opening such as a door or window.

Lookout—A short wood bracket or cantilever to support an overhanging portion of a roof or the like, usually concealed from view.

Louver—An opening with a series of horizontal slats so arranged as to permit ventilation but to exclude rain, sunlight, or vision. See also *Attic ventilators*.

Lumber—Lumber is the product of the sawmill and planing mill not further manufactured other than by sawing, resawing, and passing lengthwise through a standard planing machine, crosscut to length, and matched.

Lumber, boards—Yard lumber less than 2 inches thick and 2 or more inches wide.

Lumber, dimension—Yard lumber from 2 inches to, but not including, 5 inches thick, and 2 or more inches wide. Includes joists, rafters, studding, planks, and small timbers.

Lumber, dressed size—The dimensions of lumber after shrinking from the green dimension and after planing, usually $1/2$ inch less than the nominal or rough size. For example, a 2×4 stud actually measures $1\frac{1}{2}$ by $3\frac{1}{2}$ inches.

Lumber, matched—Lumber that is edge-dressed and shaped to make a close tongue-and-groove joint at the edges or ends when laid edge to edge or end to end.

Lumber, shiplap—Lumber that is edge-dressed to make a close rabbeted or lapped joint.

Lumber, timbers—Lumber 5 or more inches in the least dimension. Includes beams, stringers, posts, caps, sills, girders, and purlins.

Mantel—The shelf above a fireplace. Originally referred to the beam or lintel supporting the arch above the fireplace opening. Used also in referring to the entire finish around a fireplace, covering the chimney breast across the front and sometimes on the sides.

Masonry—Stone, brick, concrete, hollow tile, concrete block, gypsum block, or other similar building units or materials or a combination of the same, bonded together with mortar to form a wall, pier, buttress, or similar mass.

Metal lath—Sheets of metal that are slit and drawn out to form openings on which plaster is spread.

Millwork—Generally, all building materials made of finished wood and manufactured in millwork plants and planing mills are included under the term *millwork*. It includes such items as inside and outside doors, window and doorframes, blinds, porchwork, mantels, panelwork, stairways, moldings, and interior trim. It does not include flooring, ceiling, or siding.

Miter—The joining of two pieces at an angle that bisects the angle of junction.

Moisture content of wood—Weight of the water contained in the wood, usually expressed as a percentage of the weight of the oven-dry wood.

Mortise—A slot cut into a board, plank, or timber, usually edgewise, to receive the tenon of another board, plank, or timber to form a joint.

Molding—Material, usually patterned strips, used to provide ornamental variation of outline or contour, whether projections or cavities (such as cornices, bases, window and doorjambs, and heads).

Mullion—A slender bar or pier forming a division between panels or units of windows, screens, or similar frames.

Muntin—The members dividing the glass or openings of sash, doors, and the like.

Natural finish—A transparent finish (usually a drying oil, sealer, or varnish) applied on wood for the purpose of protection against soiling or weathering. Such a finish may not seriously alter the original color of the wood or obscure its grain pattern.

Newel—Any post to which a stair railing or balustrade is fastened.

Nonbearing wall—A wall supporting no load other than its own weight.

Nosing—The projection edge of a molding or drip. Usually applied to the projecting molding on the edge of a stair tread.

o. c. (on center)—The measurement of spacing for studs, rafters, joists, and the like in a building from the center of one member to the center of the next member.

o. g. (ogee)—A molding with a profile in the form of a letter S, having the outline of a reversed curve.

Paint—L, pure white lead (basic-carbonate) paint; TLZ, titanium-lead-zinc paint; TZ, titanium-zinc paint.

Panel—A large, thin board or sheet of lumber, plywood, or other material. A thin board with all its edges inserted in a groove of a surrounding frame of thick material. A portion of a flat surface recessed or sunk below the surrounding area, distinctly set off by molding or some other decorative device. Also, a section of floor, wall, ceiling, or roof, usually prefabricated and of large size, handled as a single unit in the operations of assembly and erection.

Paper, building—A general term for papers, felts, and similar sheet materials used in buildings without reference to their properties or uses.

Paper, sheathing—A building material (generally paper or felt) used in wall and roof construction as a protection against the passage of air and sometimes moisture.

Parting stop or strip—A small wood piece used in the side and head jambs of double-hung windows to separate the upper and lower sash.

Partition—A wall that subdivides spaces within any story of a building.

Penny—As applied to nails, it originally indicated the price per hundred. The term now serves as a measure of nail length and is abbreviated by the letter "d."

Pier—A column of masonry, usually rectangular in horizontal cross-section, used to support other structural members.

Pigment—A powdered solid in suitable degree of subdivision for use in paint or enamel.

Pitch—The incline or rise of a roof. Pitch is expressed in inches or rise per foot of run, or by the ratio of the rise to the span.

Pitch pocket—An opening extending parallel to the annual rings of growth that usually contains, or has contained, either solid or liquid pitch.

Pith—The small, soft core at the original center of a tree around which wood formation takes place.

Plate—(*a*) A horizontal structural member placed on a wall or supported on posts, studs, or corbels to carry the trusses of a roof or to carry the rafters directly. (*b*) A shoe, or base member, as of a partition or other frame. (*c*) A small, relatively flat member placed on or in a wall to support girders, rafters, and so on.

Plough—To cut a groove, as in a plank.

Plumb—Exactly perpendicular (vertical).

Ply—A term to denote the number of thicknesses or layers of roofing felt, veneer in plywood, or layers in built-up materials in any finished piece of such material.

Plywood—A piece of wood made of three or more layers of veneer joined with glue and usually laid with the grain of adjoining plies at right angles. Almost always an odd number of plies are used to provide balanced construction.

Porch—A floor extending beyond the exterior walls of a building. It may be covered and enclosed or unenclosed.

Pores—Wood cells of comparatively large diameter that have open ends and are set one above the other to form continuous tubes. The openings of the vessels on the surface of a piece of wood are referred to as *pores*.

Preservative—Any substance that, for a reasonable length of time, will prevent the action of wood-destroying fungi, borers of various kinds, and similar destructive life when the wood has been properly coated or impregnated with it.

Primer—The first coat of paint in a paint job that consists of two or more coats. Also, the paint used for such a first coat.

Putty—A type of cement usually made of whiting and boiled linseed oil, beaten or kneaded to the consistency of dough and formerly used in sealing glass in sash. Today, glazing compound is used instead.

Quarter round—A molding that presents a profile of a quarter circle.

Rabbet—A rectangular longitudinal groove cut in the corner of a board or other piece of material.

Radiant heating—A method of heating, usually consisting of coils or pipes placed in the floor, wall, or ceiling.

Rafter—One of a series of structural members of a roof designed to support roof loads. The rafters of a flat roof are sometimes called *roof joists*.

Rafter, hip—A rafter that forms the intersection of an external roof angle.

Rafter, jack—A rafter that spans the distance from a wallplate to a hip or from a valley to a ridge.

Rafter, valley—A rafter that forms the intersection of an internal roof angle.

Rail—A horizontal bar or timber of wood or metal extending from one post or support to another as a guard or barrier in a fence, balustrade, staircase, and so on. Also, the cross- or horizontal members of the framework of a sash, door, blind, or any paneled assembly.

Rake—The trim members that run parallel to the roof slope and from the finish between wall and roof.

Raw linseed oil—The crude product expressed from flaxseed and usually without much subsequent treatment.

Reinforcing—Steel rods or metal fabric placed in concrete slabs, beams, or columns to increase their strength.

Relative humidity—The amount of water vapor expressed as a percentage of the maximum quantity that could be present in the atmosphere at a given temperature. The actual amount of water vapor that can be held in space increases with the temperature.

Resin-emulsion paint—Paint, the vehicle (liquid part) of which consists of resin or varnish dispersed in fine droplets in water, analogous to cream (which is butterfat dispersed in water).

Ribbon—A narrow board let into the studding to add support to joists.

Ridge—The horizontal line at the junction of the top edges of two sloping roof surfaces. The rafters are nailed at the ridge.

Ridge board—The board placed on edge at the ridge of the roof to support the upper ends of the rafters.

Rise—The height of a roof rising in horizontal distance (run) from the outside face of a wall supporting the rafters or trusses to the ridge of the roof. In stairs, the perpendicular height of a step or flight of steps.

Riser—Each of the vertical boards closing the spaces between the treads of the stairways.

Roll roofing—Roofing material, composed of fiber and saturated with asphalt, that is supplied in rolls containing 108 square feet in 36-inch widths. It is generally furnished in weights of 55 to 90 pounds per roll.

Roof sheathing—The boards or sheet material fastened to the roof rafters on which the shingles or other roof covering is laid.

Rubber-emulsion paint—Paint, the vehicle of which consists of rubber or synthetic rubber dispersed in fine droplets in water.

Run—In reference to roofs, the horizontal distance from the face of a wall to the ridge of the roof. Referring to stairways, the net width of a stop. Also, the horizontal distance covered by a flight of steps.

Sapwood—The outer zone of wood, next to the bark. In the living tree it contains some living cells (the heartwood contains none), as well as dead and dying cells. In most species, it is lighter colored than the heartwood. In all species, it is lacking in decay resistance.

Sash—A single frame containing one or more panes of glass.

Sash balance—A device, usually operated with a spring, designed to counterbalance window sash. Use of sash balances eliminates the need for sash weights, pulleys, and sash cord.

Saturated felt—A felt impregnated with tar or asphalt.

Scratch coat—The first coat of plaster, which is scratched to form a bond for the second coat.

Scribing—Fitting woodwork to an irregular surface.

Sealer—A finishing material, either clear or pigmented, that is usually applied directly over uncoated wood for the purpose of sealing the surface.

Seasoning—Removing moisture from green wood to improve its serviceability.

Semi-gloss paint or enamel—A paint or enamel made with a slight insufficiency of nonvolatile vehicle so that its coating, when dry, has some luster, but is not highly glossy.

Shake—A handsplit shingle, usually edge-grained.

Sheathing—The structural covering (usually wood boards, plywood, or wallboards) placed over exterior studding or rafters of a structure.

Sheathing paper—See *Paper, sheathing*.

Shellac—A transparent coating made by dissolving lac, a resinous secretion of the lac bug (a scale insect that thrives in tropical countries, especially India), in alcohol.

Shingles—Roof covering of asphalt, asbestos, wood, tile, slate, or other material cut to stock lengths, widths, and thicknesses.

Shingles, siding—Various kinds of shingles, some specially designed, that can be used as the exterior side-wall covering for a structure.

Shiplap—See *Lumber, shiplap*.

Siding—The finish covering of the outside wall of a frame building, whether made of weatherboards, vertical boards with battens, shingles, or other material.

Siding, bevel (lap siding)—Used as the finish siding on the exterior of a house or other structure. It is usually manufactured by resawing dry square-surfaced boards diagonally to produce two wedge-shaped pieces. These pieces commonly run from $3/16$ inch thick on the thin edge to $1/2$ to $3/4$ inch thick on the other edge, depending on the width of the siding.

Siding, drop—Usually $3/4$ inch thick and 6 inches wide, machined into various patterns. Drop siding has tongue-and-groove joints, is heavier, has more structural strength, and is frequently used on buildings that require no sheathing (such as garages and barns).

Sill—The lowest member of the frame of a structure, resting on the foundation and supporting the uprights of the frame. The member forming the lower side of an opening, as a door sill, window sill, etc.

Soffit—The underside of the members of a building, such as staircases, cornices, beams, and arches, relatively minor in area as compared with ceilings.

Soil cover (ground cover)—A light roll roofing or plastic used on the ground of crawl spaces to minimize moisture permeation of the area.

Soil stack—A general term for the vertical main of a system of soil, waste, or vent piping.

Sole or soleplate—A member, usually a 2 × 4, on which wall and partition studs rest.

Span—The distance between structural supports (such as walls, columns, piers, beams, girders, and trusses).

Splash block—A small masonry block laid with the top close to the ground surface to receive roof drainage and to carry it away from the building.

Square—A unit of measure (100 square feet) usually applied to roofing material. Side-wall coverings are often packed to cover 100 square feet and are sold on that basis.

Stain, shingle—A form of oil paint, very thin in consistency, intended for coloring wood with rough surfaces (such as shingles), but without forming a coating of significant thickness or with any gloss.

Stair carriage—A stringer for steps on stairs.

Stair landing—A platform between flights of stairs or at the termination of a flight of stairs.

Stair rise—The vertical distance from the top of one stair tread to the top of the one next above.

Stool—The flat, narrow shelf forming the top member of the interior trim at the bottom of a window.

Storm sash or storm window—An extra window usually placed on the outside of an existing window as additional protection against cold weather.

Story—That part of a building between floors and the floor or roof above.

String, stringer—A timber or other support for cross members. In stairs, the support on which the stair treads rest. Also called a *stringboard.*

Stucco—Most commonly refers to an outside plaster made with Portland cement as its base.

Stud—One of a series of slender wood or metal structural members placed as supporting elements in walls and partitions. (Plural: studs or studding.)

Subfloor—Boards or sheet material laid on joists over which a finish floor is to be laid.

Tail beam—A relatively short beam or joist supported in a wall on one end and by a header on the other.

Termite—Insects that superficially resemble ants in size, general appearance, and habit of living in colonies. Hence, they are frequently called *white ants*. Subterranean termites do not establish themselves in buildings by being carried in with lumber, but by entering from ground nests after the building has been constructed. If unmolested, they eat out the woodwork, leaving a shell of sound wood to conceal their activities, and damage may proceed so far as to cause collapse of parts of a structure before discovery. There are about 56 species of termites known in the United States. However, the two major species, classified from the manner in which they attack wood, are ground-inhabiting or subterranean termites (the most common) and dry-wood termites (found almost exclusively along the extreme southern border and the Gulf of Mexico in the United States).

Termite shield—A shield, usually of noncorrodible metal, placed in or on a foundation wall or other mass of masonry or around pipes to prevent passage of termites.

Threshold—A strip of wood or metal beveled on each edge and used above the finished floor under outside doors.

Toenailing—To drive a nail at a slant with the initial surface to permit it to penetrate into a second member.

Tread—The horizontal board in a stairway on which the foot is placed.

Trim—The finish materials in a building, such as moldings, applied around openings (window trim, door trim) or at the floor and ceiling of rooms (baseboard, cornice, picture molding).

Trimmer—A beam or joist to which a header is nailed in framing for a chimney, stairway, or other opening.

Truss—A frame or jointed structure designed to act as a beam of long span, while each member is usually subjected to longitudinal stress only, either tension or compression.

Trisodium phosphate (TSP)—Used to clean wood or masonry before painting. Usually obtainable from hardware stores. Often used by power washers to quickly and thoroughly clean surfaces from fences, decks, and wood-sided houses before repainting.

Turpentine—A volatile oil used as a thinner in paints and as a solvent in varnishes. Chemically, it is a mixture of terpenes.

Undercoat—A coating applied prior to the finishing or top coats of a paint job. It may be the first of two or the second of three coats. In some usage of the word, it may become synonymous with priming coat.

Valley—The internal angle formed by the junction of two sloping sides of a roof.

Vapor barrier—Material used to retard the flow of vapor or moisture into walls and thus to prevent condensation within them. There are two types of vapor barriers: the membrane (which comes in rolls and is applied as a unit in the wall or ceiling construction) and the paint type (which is applied with a brush). The vapor barrier must be a part of the warm side of the wall.

Varnish—A thickened preparation of drying oil, or drying oil and resin, suitable for spreading on surfaces to form continuous, transparent coatings, or for mixing with pigments to make enamels.

Vehicle—A liquid portion of a finishing material. It consists of the binder (nonvolatile) and volatile thinners.

Veneer—Thin sheets of wood.

Vent—A pipe installed to provide a flow of air to or from a drainage system or to provide a circulation of air within such systems to protect trap seals from siphonage and back pressure.

Vermiculite—A mineral closely related to mica, with the faculty of expanding on heating to form lightweight material with insulation quality. Used as bulk insulation and as aggregate in insulating and acoustical plaster, as well as in insulating concrete floors.

Volatile thinner—A liquid that evaporates readily and is used to thin or reduce the consistency of finishes without altering the relative volumes of pigments and nonvolatile vehicle.

Wallboard—Wood pulp, gypsum, or other materials made into large rigid sheets that may be fastened to the frame of a building to provide a surface finish.

Wane—Bark, or lack of wood or bark from any cause, on the edge or corner of a piece.

Wash—The upper surface of a member or material when given a slope to shed water.

Water repellent—A liquid designed to penetrate into wood and to impart water repellency to the wood.

Water table—A ledge or offset on or above a foundation wall, for the purpose of shedding water.

Weatherstrip—Narrow strips made of metal, or other material, so designed that when installed at doors or windows they will retard the passage of air, water, moisture, or dust around the door or window sash.

Wood rays—Strips of cells extending radially within a tree and varying in height from a few cells in some species to 4 inches or more in oak. The rays serve primarily to store food and to transport it horizontally in the tree.

Index

P